BURLEIGH DODDS SCIENCE: INSTANT INSIGHTS

NUMBER 98

Improving the sustainability of dairy production

Published by Burleigh Dodds Science Publishing Limited
82 High Street, Sawston, Cambridge CB22 3HJ, UK
www.bdspublishing.com

Burleigh Dodds Science Publishing, 1518 Walnut Street, Suite 900, Philadelphia, PA 19102-3406, USA

First published 2024 by Burleigh Dodds Science Publishing Limited
© Burleigh Dodds Science Publishing, 2024, except the following: the contribution of f Dr Stephanie A. Terry, Dr Carlos M. Romero, Dr Alex V. Chaves and Dr Tim A. McAllister in Chapter 3 is © Her Majesty the Queen in Right of Canada. All rights reserved.

British Library Cataloguing in Publication Data
A catalogue record for this book is available from the British Library

ISBN 978-1-80146-669-1 (Print)
ISBN 978-1-80146-670-7 (ePub)

DOI: 10.19103/9781801466707

Typeset by Deanta Global Publishing Services, Dublin, Ireland

Contents

Series list

Title	Series number
Sweetpotato	01
Fusarium in cereals	02
Vertical farming in horticulture	03
Nutraceuticals in fruit and vegetables	04
Climate change, insect pests and invasive species	05
Metabolic disorders in dairy cattle	06
Mastitis in dairy cattle	07
Heat stress in dairy cattle	08
African swine fever	09
Pesticide residues in agriculture	10
Fruit losses and waste	11
Improving crop nutrient use efficiency	12
Antibiotics in poultry production	13
Bone health in poultry	14
Feather-pecking in poultry	15
Environmental impact of livestock production	16
Sensor technologies in livestock monitoring	17
Improving piglet welfare	18
Crop biofortification	19
Crop rotations	20
Cover crops	21
Plant growth-promoting rhizobacteria	22
Arbuscular mycorrhizal fungi	23
Nematode pests in agriculture	24
Drought-resistant crops	25
Advances in detecting and forecasting crop pests and diseases	26
Mycotoxin detection and control	27
Mite pests in agriculture	28
Supporting cereal production in sub-Saharan Africa	29
Lameness in dairy cattle	30
Infertility and other reproductive disorders in dairy cattle	31
Alternatives to antibiotics in pig production	32
Integrated crop–livestock systems	33
Genetic modification of crops	34

Acknowledgements

Chapters in this Instant Insight are taken from the following sources:

Chapter 1 Setting environmental targets for dairy farming
   Chapter taken from: van Belzen, N. (ed.), Achieving sustainable production of milk Volume 2: Safety, quality and sustainability, Burleigh Dodds Science Publishing, Cambridge, UK, 2017, (ISBN: 978 1 78676 048 7)

Chapter 2 Welfare of pigs during finishing
   Chapter taken from: van Belzen, N. (ed.), Achieving sustainable production of milk Volume 2: Safety, quality and sustainability, Burleigh Dodds Science Publishing, Cambridge, UK, 2017, (ISBN: 978 1 78676 048 7)

Chapter 3 Nutritional factors affecting GHG emissions from ruminants
   Chapter taken from: McSweeney, C. S. and Mackie, R. I. (ed.), Improving rumen function, Burleigh Dodds Science Publishing, Cambridge, UK, 2019, (ISBN: 978 1 78676 332 7)

Chapter 4 Host-rumen microbiome interactions and influences on feed conversion efficiency (FCE)
   Chapter taken from: McSweeney, C. S. and Mackie, R. I. (ed.), Improving rumen function, Burleigh Dodds Science Publishing, Cambridge, UK, 2019, (ISBN: 978 1 78676 332 7)

Chapter 5 Developing closed-loop dairy value chains and tools to support decision-makers
   Chapter taken from: Amon, B. (ed.), Developing circular agricultural production systems, pp. 369-400, Burleigh Dodds Science Publishing, Cambridge, UK, 2024, (ISBN: 978 1 80146 256 3)

Chapter 1

Setting environmental targets for dairy farming

Sophie Bertrand, French Dairy Inter-branch Organization, France

1 Introduction

The demand in milk production is increasing with the growing world population, but the world's natural resources are limited. This is why it is essential for the dairy sector to produce more milk but in a more sustainable way, as we cannot add to the pressure on the environment which is already coping with climate change.

Setting environmental targets for dairy farming requires an ability to manage the environmental impact of dairy farming, which in turn requires an ability to measure those impacts. However, it can be very difficult for evaluating the environmental impact of dairy farming, as dairy systems are so varied and operate in such diverse climates and geographical contexts. To add to the complexity, environmental impacts are all interlinked, so that taking action to reduce one without considering the consequences on the others can be dangerous and lead to wrong decisions. In addition, science is not sufficiently advanced to be able to measure all the impacts with the same precision: positive impacts are very difficult to evaluate and therefore are often ignored.

http://dx.doi.org/10.19103/AS.2016.0005.28

This chapter reviews the main international methods available to evaluate the environmental impact of dairy farming, identify levers of action and set environmental targets. It has to be kept in mind that with such diverse production situations, environmental impact and possible intervention strategies, any global environmental assessment is a simplification of reality.

2 A global typology of dairy production systems for use in environmental assessments

One of the characteristics of worldwide dairy production is its diversity, which is a strength as each system is adapted to its specific context, but which also makes it very difficult to evaluate the global impact on the environment. To address the diversity of dairy systems and of geographical and climate contexts, in 1996 the FAO developed (Seré and Steinfeld, 1996) a global typology of dairy production systems. This proposed a very schematic production system typology based on two major feed-base system types: mixed- and grass-based, classified into three major agro-ecological zones: temperate regions, arid and semi-arid tropics, and sub-humid and humid tropics. This typology is used by the FAO to evaluate the environmental impact of dairy farming at the world level with the tool GLEAM (Global Livestock Environmental Assessment Model). Global Livestock Environmental Assessment Model-interactive (GLEAMi) is a modelling framework that simulates the interaction of activities and processes involved in livestock production and the environment. It was developed by the FAO to assess livestock impacts, adaptation and mitigation options at (sub) national, regional and global levels. It is an Excel-based publicly available tool. It can be used to support the preparation of greenhouse gas inventories, project design and evaluation, as well as analytical work on emissions, but it cannot be used at the farm level. GLEAMi is based on Intergovernmental Panel on Climate Change (IPCC) Tier 2 (IPCC, 2006) computation guidelines and benefits from the in-built GLEAM database of parameters and activity data, which can be adapted to match specific conditions. The tool can be downloaded at: https://bit.ly/resources_GLEAM. The methodology used in this tool is based on LCA.

3 Life cycle assessment (LCA): an overview

Life cycle assessment (LCA) has become the internationally agreed method to address the complexity of interlinked and multiple impacts in food production. LCA gives comprehensive environmental information to help identify environmentally sustainable agricultural products and practices. ISO 14040:2006 (ISO, 2006) defines LCA as a 'compilation and evaluation of the inputs, outputs and the potential environmental impacts of a product system throughout its life cycle'.

The LCA approach, which is defined in ISO standards 14040 and 14044 (ISO, 2006), was originally used for industrial processes, but is now widely used in agriculture as a method for evaluating the environmental impact of production and identifying the resource and emission-intensive processes within a product's life cycle. The main strength of LCA lies in its ability to provide a holistic assessment of resource use and environmental impacts during the life cycle of a product (ISO, 2006). LCA helps identify effective approaches to reduce environmental burdens and evaluate the effect that changes within a production process may have on the overall life cycle balance of environmental burdens. This enables the identification and exclusion of measures that simply shift environmental problems from one phase of the life cycle to another. It involves the systemic analysis of production systems accounting for all inputs and outputs associated with a specific product within a defined 'system boundary'. The system boundaries determine which unit processes are included in the LCA study. Defining system boundaries is partly based on subjective choice made during the scope phase when the boundaries are initially set. The system boundary depends on the aim of the study. The 'reference unit' that is the useful output of the production system is the functional unit, and it has a defined quantity and quality. The functional unit can be based on a defined quantity, such as 1 kg of product, or it may be based on an attribute of a product or process, such as 1 kg of fat and protein corrected milk.

Attributional LCA is more commonly used. It estimates the environmental burden of the existing situation under current production and market conditions and allocates impacts to the various co-products of the production system. This differs from the consequential LCA approach, which considers potential consequences of changes in production and relies on a system expansion analysis to allocate impacts of co-products (Thomassen et al., 2008). When a system has more than one product, ISO recommends avoiding partitioning the system but instead expanding the product system to include the additional functions related to the co-products, such as meat. However, this has rarely been used in the case of dairy production.

LCA still presents significant challenges and limits when applied to agriculture. First, the method is data-intensive, which is a problem with biological systems (e.g. soil and climate), where data is difficult to collect. Limited data availability can result in simplifications, which can in turn lead to inaccuracies and a high level of uncertainty. A second difficulty lies in the fact that methodological choices are still possible when following the ISO guidelines, such as system boundary, functional units and allocation, which can make a big difference to the results, even with the same initial data. The main methodological choice that has an impact on the result is the allocation method. The application of LCA to agricultural systems is complicated because major products are usually accompanied by the joint production of by-products, such as meat and milk in the case of dairy farming and oil and soya cake in the case of soya. This requires partitioning of environmental impacts to each product from the system according to an allocation rule, which may be based on different criteria such as economic value, mass balances, product properties and so on. Careful analysis shows significant differences between results when using a different allocation method.

That is why the FAO and IDF have developed sectorial guidelines reducing the methodological choices to allow comparison of results from different LCA studies realized on dairy products. The international dairy sector proactively wrote ISO-based harmonized guidelines to produce results that are comparable at world level (Bertrand et al., 2010). Aware of these methodological challenges, and to help improve the environmental performance of the livestock supply chain, in 2012 the FAO launched the Livestock Environmental Assessment and Performance (LEAP). LEAP provides a platform for the harmonization of metrics and methods to monitor the environmental performance of the livestock supply chains. The partnership develops broadly recognized sector-specific guidelines and metrics for assessing and monitoring the environmental performance of the livestock sector. The benefit of a sector-specific approach is that it gives guidance on the application of LCA to users in line with the ISO guidance and provides a common basis from which to evaluate resource use and environmental impacts. In 2014 and 2015, LEAP published four guides: LCA of feed supply chain (LEAP, 2014), LCA of large ruminant products, LCA of small ruminant products and biodiversity assessment guidelines (LEAP, 2015). An LCA database on feed is also available on the LEAP website to help the users in LCA calculation. Information and documents can be found at: http://www.fao.org/partnerships/leap/en/.

In coordination with the FAO, the IDF also developed dairy sector-specific guidelines that are more detailed and easier to use for the dairy industry.

4 LCA: product carbon footprint

4.1 A common carbon footprint method

The product carbon footprint is the outcome of the analysis of greenhouse gas emissions (GHG) throughout the life cycle of a product in relation to a defined functional unit. GHG emissions comprise all gaseous substances for which the IPCC has defined a global warming potential coefficient and which are to be expressed in mass-based CO_2 equivalents. A product carbon footprint is based on an LCA methodology. The main agricultural greenhouse gases are carbon dioxide, nitrous oxide and methane (ISO/TS 14067:2013)

In 2010, the IDF managed to reach a consensus on a common methodology for carbon footprinting (IDF, 2010), which was an important step to improve the understanding of emission sources and possible mitigation (Bertrand et al., 2010). This harmonized method allowed dairy stakeholders to identify a 20% reduction potential within the same type of dairy system. Scientists calculated GHGs on a significant number of dairy farms with different types of dairy systems (536 farms in the US dairy sector study and 3900 farms in the French dairy sector study) (Thoma et al., 2013). The results show a difference of 20% between the emissions of the 'low carbon' dairy producers compared to the 'high carbon' producers within the same type of production system. The difference in results was explained by very technical levers of action, linked with the technical management of the farm (e.g. fertilizer management, type and quantity of feed, and herd management). This understanding of the magnitude, sources and pathways of GHG emission was essential to avoid oversimplification in messages and action. It helped the dairy sector make more informed choices. In 2013 the FAO published the technical report on GHG emission from the ruminant supply chain. This report provided a comprehensive and disaggregated global assessment of emissions from the ruminant sector based on a common methodology, which enabled the understanding of main emission pathways and hot spots. This was a fundamental first step towards identification of mitigation strategies (Opio et al., 2013).

In 'tackling climate change through livestock' (Gerber et al. 2013), the FAO identified ways to reduce emissions by assessing the mitigation potential of different sets of technologies. The results of the estimation on the dairy sector showed that if producers applied the practices of the 25% of producers with lowest emission intensity (25th percentile), the sector's emissions could be reduced by 18% (about 1.1 gigatonnes CO_2-eq). A 30% reduction of GHG emission would be possible if producers in a given system, region and climate adopted the practices used by the 10% of producers with the lowest emission intensity.

An important aspect of the dairy sector is that it can store carbon in the soil of pastures used by dairy cows (Soussana et al., 2007), which is also a challenge as no international method is available to estimate this carbon storage. Grasslands cover about one-quarter of the earth's land surface. On a global scale, livestock use 3.4 billion hectares of grazing land (i.e. grasslands and rangelands) in addition to animal feed produced on about a quarter of the land under crops. Agricultural ecosystems hold large C reserves in organic soil matter, which is recognized by the IPCC (IPCC, 2001). Soil C sequestration is the mechanism responsible for most of the mitigation potential in the agriculture sector, with an estimated 89% contribution compared to 9% and 2% for mitigation of methane and nitrous oxide emissions, respectively (IPCC, 2007). Worldwide, the soil organic carbon (SOC) sequestration potential is estimated to be 0.01-0.3 Gt C/year on 3.7 billion ha of permanent pasture (Soussana et al., 2009). Carbon accumulation in grassland ecosystems occurs mostly below ground. Grassland soils are typically rich in SOC, owing partly to active rhizodeposition and partly to the activity of earthworms.

Rhizodeposition favours C storage because direct incorporation into the soil matrix allows a high degree of physical stabilization of the organic soil matter. While there has been a steady C accumulation in the soils of many ecosystems over millennia, it is usually thought that soil C accumulation capacity is limited and that after several centuries old non-disturbed soils should have reached equilibrium in terms of their C balance. However, soil C sequestration is reversible, as factors such as soil disturbance, vegetation degradation, fire, erosion, nutrient shortage and water deficit may all lead to a rapid loss of SOC. Therefore, the carbon sequestration in grasslands has huge potential in mitigation of emissions, but a lot more research and development is required to develop measurement and evaluation methodologies.

4.2 Case study: the French low carbon dairy farm initiative

The IDF harmonized LCA method allowed stakeholders, like the French dairy sector, to implement effective action on dairy farms. The French Livestock Institute developed a tool, CAP'2ER, based on the harmonized LCA method, which can measure the GHG emissions at the dairy farm level, identify the levers of action and simulate the effect of the changes proposed on the technical, environmental and economic performance of the farm. CAP'2ER stands for 'Calcul Automatisé des Performances Environnementales en Elevage de Ruminants'. It is meant to calculate environmental performance (e.g. carbon stocks, GHG, eutrophication, acidification, energy and biodiversity) of ruminants. There are two versions of the tool: a simple education/awareness version online, and a decision-making version to be used by farm advisers. With such a helpful tool, it is possible to transfer knowledge between farmers on levers of action on a large scale. Consequently, in 2015, the French dairy sector launched an ambitious carbon road map, 'the low carbon dairy farm', involving all the dairy stakeholders at the national level. Farm advisers were trained to use the CAP'2ER tool and went to visit more than 5000 volunteer dairy farmers in 2015 and 2016. The use of the tool helped the 5000 farmers identify the specific levers of action on their own farm and build an action plan to reduce emissions at the farm level. With this national initiative involving the main stakeholders of the dairy chain, the ambition of the French dairy sector is to reduce the carbon footprint of French milk by 20% in 2025. More information can be found at: www.low-carbon-dairy-farm.com

A similar approach has been developed by the Innovation Center for US Dairy, with the tool 'Farm Smart', based on the same common methodology.

5 LCA: product water footprint

The water footprint (WF) is the fraction of an LCA that is related to water. It includes impacts associated with water use (quantity) and the effect on water availability for humans and ecosystems, as well as direct impacts on water resources and its users from emissions to air and soil and reduced water quality. The potential impact on water quality is quantified using traditional LCA impact categories namely freshwater eutrophication, freshwater acidification, human toxicity and eco-toxicity.

Principles, requirements and guidelines for the assessment of a WF are defined in the DIS2 ISO 14046 document, 2014. The main principle highlighted in this document is that a 'water scarcity footprint' and a 'water availability footprint' assess impacts associated with water use only, whereas a WF (no qualifier) assesses all relevant impacts related to water, quantity and quality. Consequently, a WF should take into account the impact on the environment and not just the 'quantity' of water. This principle is very well explained by Ridoutt:

> Environmental relevance must be taken into consideration if water footprints are to inform decision making and policy development. Water use in a region of abundance does not have the same potential to impact human wellbeing and ecosystem health as water use in a region of water scarcity. The need to reduce humanity's water footprint does not arise from an absolute spatial and temporal shortage of freshwater in the world. It is the result of the current pattern of freshwater use, which is greatly skewed toward highly stressed watersheds. Environmental relevance is the key to understanding water footprints (Ridoutt et al., 2012).

As explained previously, LCA can be used to assess the environmental impact throughout a product's life cycle. But unlike carbon footprinting, which is now a straight-forward documented procedure based on computing global warming potentials within the LCA, a range of methods for estimating water consumption have been developed. The main water-use assessment methods applied to agricultural products are:

- The WF from the WaterFootprintNetwork, which is an indicator of freshwater consumption developed by water management scientists. WF takes the direct and indirect water consumption of a product into account (blue and green water). Thereafter, the sustainability component of the WF within a catchment or river basin can be assessed.
- LCA methods using the ISO standardized framework (ISO 14046, 2010). One LCA method is based on assessing the midpoint indicator 'water deprivation', as a function of freshwater scarcity considering hydrological conditions (Ridoutt and Pfister, 2010; Zonderland-Thomassen, 2012). A second LCA method is based on assessing freshwater use impacts related to the availability of sufficient freshwater for ecosystems, freshwater depletion and changes in the water cycle caused by land use change (Boulay 2015) (http://wulca-waterlca.org/).

So far no consensus has been reached within the water scientist community, which is why the ISO document gives only principles but does not recommend any method to assess the WF of a product.

Applying different water-use assessment methods not only complicates comparison of results between products and countries, but also complicates communication to stakeholders and identification of potential mitigation strategies.

In 2016 the FAO launched a project within LEAP to reach an agreed method to assess the WF of livestock products. While the work is still ongoing at the FAO, the IDF guide to water footprint methodology for the Dairy sector (IDF, 2017). The science is less advanced on this water issue than on carbon, so the dairy stakeholders have not yet estimated potential targets and main levers of action.

6 Assessing impacts on biodiversity

6.1 Using the right method to assess biodiversity impacts

Biodiversity is defined as 'Variability among living organisms from all sources including inter alia, terrestrial, marine and other aquatic systems and ecological complexes of which they are part including diversity within species and of ecosystems' (Article 2 of the Convention of Biological Diversity). The assessment of livestock impacts on biodiversity is an emerging area of work.

Biodiversity is complex, multivariate by nature and context-dependent, which in turn makes the assessment of dairy production impacts on biodiversity extremely complicated. The main principle that should be highlighted is that effects of dairy production on biodiversity can be both positive and negative, depending on the context and practices; consequently assessment methods need to be capable of reflecting beneficial as well as detrimental impacts. Another important principle on this biodiversity issue is that the choice of reference state used as the baseline for comparisons plays an important role in the assessment, and the results should be interpreted accordingly. Due to the complexity of the biodiversity issue, it is only possible today to use broad consensual principles to identify context-specific biodiversity indicators.

In 2015, the FAO published a synthesis on livestock and biodiversity (LEAP, 2015). The FAO report indicates that the LCA methodology is not the most appropriate to assess biodiversity with the knowledge available today. The pressure, state, response (PSR) approach is more accurate for biodiversity. This approach can be applied to identify biodiversity impacts with a higher level of detail. Indicators are defined at the level of pressures generated by anthropogenic activities, which lead to a change in the environment state. This leads to activities of stakeholders in order to reach a more sustainable situation as the response. Therefore, most biodiversity initiatives in the livestock sector today rely on PSR indicators and response indicators in particular, as opposed to LCA approaches. As noted in the LEAP *Principles for the assessment of livestock impacts on biodiversity* , 2015, the 'LCA approach for biodiversity assessment needs substantial improvements as it is unable to grasp the real and complex dynamics of ecosystem interactions'. However, off-farm feed production, national or international, should be included in the study to avoid a shift of pollution. The idea is to try and include as far as possible in the PSR approach, the impact on biodiversity related to the cultivation of the feed purchased off farm, without following a real LCA. This task is difficult as very often the information (data on biodiversity impact) just does not exist when feed is produced in a different continent.

Because of this high complexity, the international dairy sector was unable to reach a consensus on a single common method to assess the impact of dairy production on biodiversity, but rather an approach that will follow steps such as defining the goal, realizing a hot spot analysis, setting the boundaries, setting the reference state, engaging with stakeholders and prioritizing indicators. Only the approach can be common, but not the method or the indicators as they are too specific to a region and dairy system.

6.2 *The need to recognize ecosystem services*

The LCA is a methodology that only accounts for negative impacts on the environment; however, to ensure a holistic approach it is essential to also consider the positive impacts. Dairy farming provides services to the environment and society. Environmental services are biophysical processes providing benefits, such as biodiversity maintenance, climate regulation and water purification. Cultural services are recreational, aesthetic and heritage benefits provided by livestock farming. Social services account for the contribution of livestock farming to rural community vitality and employment. Dairy production is an essential activity in many rural areas. It contributes to the maintenance of soil fertility, farmland biodiversity and provides services such as food security, rural vitality and cultural heritage.

Today, dairy sustainability assessments focus only on negative environmental impacts; thus, they are limited, because they do not consider all the benefits associated with dairy farming. The multiple contributions provided by livestock to human well-being are beginning to be acknowledged in science. However, there is no methodology available to quantify all these contributions and analyse their relationships. Indicators that evaluate social, cultural and positive environmental contributions derived from livestock farming are a lot less developed and largely underestimated or ignored altogether in sustainability assessments.

However, to be able to set environmental targets and propose action plans, it is necessary to have a comprehensive assessment of dairy farming sustainability that includes the positive social, cultural and environmental contributions. Otherwise, the risk is that decisions are taken only on the basis of negative environmental impact, without knowing what the consequences of the action will be on the positive services provided. Currently, little data is available for most services throughout the world.

The assessment of ecosystem services is an emerging area of work that is necessary if the dairy sector wants to inform stakeholders and society about the multiple benefits provided by dairy farming and help those in the livestock sector to better understand and manage these diversified services (Carrère et al., 2012).

7 Setting environmental targets: challenges and limits

Impressive progress has been made in the last 10 years in assessing the environmental impact of dairy farming. However, many limits and challenges still remain about the indicators and methods used. The methodologies presented in this chapter are the main agreed international methods to reach an initial overview of the environmental impact at a dairy farm or regional level. When it comes to defining an action plan for a farm, the studies should be complemented by tools taking into account the local context that will evaluate in detail one specific aspect of the impacts and help make decisions. Several local tools and the expertise of an adviser might be needed to be able to build an action plan with the farmer.

Another big challenge in setting environmental targets is that there is no balance between negative and positive impacts. As detailed above, indicators accounting for negative impacts are a lot more developed than indicators of positive impacts such as carbon storage or ecosystem services. This partial negative environmental assessment could lead to wrong conclusions. This limited knowledge of positive impacts should be highlighted and taken into account when writing the conclusion of an environmental assessment to avoid dangerous shortcuts.

Another limitation linked with the previous one is that surprisingly the soil is not taken into account in the methods presented above. Soil quality and fertility are essential for any production system but to date this has not been captured in global tools. The conclusion of an assessment could recommend practice changes that would deteriorate soil quality as an unintended and perhaps even undetected consequence.

Finally, it has to be kept in mind that dairy production is closely linked to meat production due to culled cows and calves. It is not possible to set environmental targets for dairy without taking into account the effect on meat production. Otherwise, the risk is to shift the burden from the dairy to the meat sector. The best example is the GHG emission from the dairy sector that can be reduced by following a strategy of dairy intensification. But by doing that, the meat production from the dairy sector decreases requiring more beef cattle to produce the meat, which then increases the GHG emission of the ruminant sector globally. As written by Puillet et al. (2014), 'In the cattle sector, milk and beef production systems are closely interconnected because of the meat obtained from dairy culled cows and fattened surplus calves. To fully evaluate a change in the dairy sector, the change in the meat production induced should also be taken into account.'

8 Conclusion

A number of excellent methods and indicators have been developed in the last ten years and they are very helpful in improving the environmental performance of the dairy sector. However, these tools have to be used with care, knowing the important limits and challenges that still remain. It also has to be kept in mind that the level of uncertainties in these assessments is still very high, as the underlying science is far from developed.

Beyond the methodological challenge, the main challenge that remains is to transfer this new knowledge to dairy farmers and stimulate an exchange between farmers and researchers, in order to capture innovation from the field and co-construct individual solutions at each farm level.

9 Where to look for further information

For further information please look at the following websites:
- www.fil-idf.org/
- dairysustainabilityframework.org/
- www.fao.org/partnerships/leap/en/
- www.low-carbon-dairy-farm.com/

10 References

Barnett, J. and Bertrand, S. 2010. Standard method for determining the carbon footprint of dairy products reduces confusion, *Animal Frontiers*.

Boulay, A. M., et al. 2015. Consensus building on the development of a stress-based indicator for LCA-based impact assessment of water consumption: outcome of the expert workshops, *Int J Life Cycle Assess*.

Carrère, P., Plantureux, S. and Pottier, E., 2012, Concilier les services rendus par les prairies pour assurer la durabilité des systèmes d'élevage herbagers (Reconciling services rendered by grassland in order to ensure the sustainability of grassland farming systems), *Fourrages*, 211, 213–18.

DIS2 ISO 14046. 2014. Environmental management – Water footprint- Principles, requirements and guidelines, *International Organization for Standardization*.

Gerber, P. J., Steinfeld, H., Henderson, B., Mottet, A., Opio, C., Dijkman, J., Falcucci, A. and Tempio, G. 2013. *Tackling Climate Change Through Livestock – A Global Assessment of Emissions And Mitigation Opportunities.* Food and Agriculture Organization of the United Nations (FAO), Rome.

IDF. 2010. Bulletin of the international dairy federation 445/2010, A common carbon footprint approach for dairy, the IDF guide to standard life cycle assessment methodology for the dairy sector, Brussels.

IPCC. 2006. IPCC guidelines for national greenhouse gas inventories. In H. S. Eggleston, L. Buendia, K. Miwa, T. Ngara and K. Tanabe (Eds), *National Greenhouse Gas Inventories Programme.* IPCC, Kanagawa.

ISO. 2006a. Environmental management – Life cycle assessment – Principles and framework. ISO 14040:2006(E). International Organization for Standardization, Geneva, Switzerland.

ISO. 2006b. Environmental management – Life cycle assessment – Requirements and guidelines. ISO 14044:2006(E). International Organization for Standardization, Geneva, Switzerland.

ISO/TS 14067. 2013. Greenhouse gases – Carbon footprint of products – Requirements and guidelines for quantification and communication.

Klumpp, K., Tallec, T., Guix, N. and Soussana, J. F. 2011. Long-term impacts of agricultural practices and climatic variability on carbon storage in a permanent pasture, *Global Change Biology*, 17, 3534–45.

LEAP. 2014. *Environmental Performance of Animal Feeds Supply Chains*, Guidelines for assessment, FAO, Rome, Italy.

LEAP. 2015. *Principles For The Assessment of Livestock Impacts On Biodiversity*, Livestock environmental assessment and performance partnership, FAO, Rome, Italy.

LEAP. 2015. *Environmental Performance of Large Ruminant Supply Chain*, Guidelines for assessment, Livestock environmental assessment and performance partnership, FAO, Rome, Italy.

Opio, C., Gerber, P., Mottet, A., Falcucci, A., Tempio, G., MacLeod, M., Vellinga, T., Henderson, B. and Steinfeld, H. 2013. *Greenhouse Gas Emissions From Ruminant Supply Chains – A Global Life Cycle Assessment.* Food and Agriculture Organization of the United Nations (FAO), Rome.

Puillet, L., et al. 2014. Modelling cattle population as lifetime trajectories driven by management options: A way to better integrate beef and milk production in emissions assessment, *Livestock Science*, http://dx. doi.org/10.1016/j.livsci.2014.04.001i.

Ridoutt, B. G. and Pfister, S. 2010. A revised approach to water footprinting to make transparent the impacts of consumption and production on global freshwater scarcity. *Global Environmental Change*, 20.

Ridoutt, et al. 2012. PNAS Letter.

Seré and Steinfeld. 1996. World livestock production systems: current status, issues and trends, Animal production and health paper No127, FAO. Rome.

Soussana, et al. 2010. Mitigating the GHG balance of ruminant production system through carbon sequestration in grasslands, *Animal*.

Thoma, et al. 2013. Regional analysis of greenhouse gas emissions from USA dairy farm: A cradle to farm gate assessment of American dairy industry circa 2008, *International Dairy Journal*. Zonderland-Thomassen, et al. 2012. Waterfootprinting – A comparison of methods using New-Zealand dairy farming as a case study, *Agricultural Systems*, 110.

Chapter 2

Improved energy and water management to minimize the environmental impact of dairy farming

J. Upton, E. Murphy and L. Shalloo, Teagasc, Ireland; M. Murphy, Cork Institute of Technology, Ireland; and I.J.M. De Boer and P.W.G. Groot Koerkamp, Wageningen University, The Netherlands

1 Introduction

We expect an increase in the demand for animal products in the future because of global human population growth (especially in developing countries), growing incomes and increasing urbanisation. The demand for animal products is expected to double by 2050 (Rae, 1998, FAO, 2009), which will create challenges to ensure that milk is produced in an environmentally efficient and sustainable manner. The livestock sector already competes increasingly for scarce resources, such as land, water and fossil energy. Reduced fossil energy use is one way to reduce greenhouse gas (GHG) emissions from dairy farming (Smith et al., 2008). Furthermore, there is increasing pressure on the dairy industry because of exposure to a globalising market in which the developments in the farm-gate price of milk are insufficient to keep pace with the increasing costs of production associated with rising energy costs (Oenema et al., 2011). Hence, energy reduction strategies are worthy of further

http://dx.doi.org/10.19103/AS.2016.0005.31

investigation as they may contribute to reductions in environmental impacts and dairy farm operating costs.

To improve the water use of an agricultural sector, insight into freshwater consumption and associated environmental impacts of individual farms is required. There is increasing recognition of the tension between livestock production and water use (Postel, 2000, Rosegrant et al., 2002, Molden et al., 2011, Busscher, 2012); hence, understanding the distribution and demands for freshwater in livestock production is of particular importance. Finite freshwater availability could become the main limiting factor for the global growth of the agri-food sector (UNEP, 2007). Quantifying the water footprint (WF) of agricultural outputs and identifying hot spots of water consumption along the food chain, therefore, is the first step in reducing the pressures on freshwater systems resulting from livestock production, while at the same time providing end-user information.

In this chapter we define a water footprint as the consumption of soil moisture due to evapotranspiration (green water) plus the consumption of groundwater and surface water (blue water), expressed as litres of water per unit output. The consumption of freshwater refers to the loss of water when it is evaporated, incorporated into a product or returned to another catchment. The distinction between green and blue water is useful for assessing and improving water use since green water dominates water use in agricultural production and globally accounts for about 80% of the water use on agricultural land (Molden et al., 2007). However, given the levels of blue water scarcity in many regions, future challenges related to total water use and water availability in agriculture will be linked to more efficient use of green water, but also increased use of green water resources (Rockström et al., 2010).

This chapter, therefore, focuses on quantification of the effect of energy reduction strategies on farm profitability, while the pillar of planet is addressed via energy depletion, exploration of solar energy, reduction of GHG emissions and quantification of water consumption in dairy production.

2 Understanding current energy use in dairy farming

2.1 Review of energy use

Energy uses on dairy farms are generally characterised as being direct or indirect (Wells, 2001). Direct energy uses are those where the energy is consumed on the farm. Examples are the use of electricity for lighting or milking and oil or diesel for crop cultivation. Indirect energy uses are those where the direct energy use occurs outside the farm boundaries. The energy use, therefore, is then embodied in the products used on the farm. Examples are energy used during the manufacture and transport of fertilisers, concentrate feed or any substantial purchases brought in for farm maintenance, such as aggregate for road maintenance (Wells, 2001).

Several studies have quantified the direct and indirect energy use (i.e. energy use up to the farm gate or along the entire life cycle) of production of dairy milk (see review of De Vries and De Boer, 2010). Fewer authors have recorded detailed inventories of the electricity consumption on commercial dairy farms. Examples of studies where electricity is measured and expressed explicitly include studies of Calcante et al. (2016), Houston et al. (2014), Wells (2001), Hartman and Sims (2006) and Basset-Mens et al. (2005).

Based on data of 150 dairy farms in New Zealand, Wells (2001) computed an average total energy use of 24.6 MJ/kg milk solids (MS), of which 38% was related to fertilisers, 21% to liquid fuels, 20% to electricity and 21% to other items. Basset-Mens et al. (2009) computed a total energy use for a national average New Zealand farm of 1.5 MJ/kg of milk. On the basis of data from 8 dairy farms in Sweden, Cederberg and Flysjö (2004) reported 2.7 MJ per kg of energy corrected milk (ECM), of which 50–60% was required for cultivation and transportation of purchased feed. On the basis of data from 119 farms in the Netherlands, Thomassen et al. (2009) reported 5.3 MJ/kg of fat and protein corrected milk (FPCM), of which 56% was required for cultivation and transport of purchased feed.

Analysis carried out by Upton et al. (2013) demonstrated that a total of 2.4 MJ was required to produce one kg of milk in a spring calving pasture-based system in Ireland, of which 20% was direct and 80% was indirect energy use (this equated to 31.7 MJ/kg of MS, 2.4 MJ/kg FPCM and 2.4 MJ/kg ECM). Electricity consumption was found to represent 12% of total cradle-to-farm-gate energy use. Other major contributors to total energy use were identified as artificial fertilisers (57% of total energy use), supplementary concentrate feed (21% of total energy use) and fuel (8% of total energy use). Electricity accounted for 60% of the direct energy use and was centred around milk harvesting. Major consumers of electricity were milk cooling (31%), water heating (23%), milking (20%), pumping water (5%) and lighting (3%), whereas other miscellaneous consumptions, such as winter housing systems, consumed 18% of the electrical energy. Hence, the electricity consumption related directly to milk harvesting (milk cooling, water heating and milking) accounted for 74% of on-farm electricity use, or about 9% of total energy use. Furthermore, over 60% of electricity was consumed at peak periods (00:00 to 09:00 h). Clearly, if peak electricity prices increase, as implied by the Energy Services Directive (EU, 2006), mitigation strategies can contribute to cost reduction.

Provisional studies in New Zealand by the New Zealand Centre for Advanced Engineering, for example, have suggested that implementation of the best management practices will produce a more energy efficient dairy farm and that the electricity consumption could decrease from 163 to 92 kWh/year per cow, representing a saving of about 44% (Morison et al., 2007). Electricity consumption in confinement systems (where cows are housed indoors and fed preserved feed and concentrate) tends to be higher than in pasture-based systems. About 865 kWh per cow per year was reported by Ludington and Johnson (2003) in a study on 32 dairy farms in New York State. Electricity consumption of tie-stall barns (934 kWh/year per cow) was higher than that of free-stall bards (811 kWh/year per cow).

2.2 Energy efficiency incentives

In the European Union, members are obliged to achieve the overall goals of the 20-20 by 2020 initiative. This initiative aims to reduce GHG emissions by 20% compared to levels in 2005, to increase the share of renewables in energy use to 20% and to improve energy efficiency by 20% by 2020 (EC, 2008). Additionally, the European Energy Services Directive 2006/32/EC was enacted to encourage improvements in energy efficiency through the implementation of improved metering of electricity coupled with incentivised demand side management (DSM) of electricity for the consumer (EU, 2006).

Smart metering implies a higher electricity price during peak periods of consumption and a lower price during off-peak periods. Peak demand on most national electricity grids is from 17:00 to 19:00 h. If dairy farmers carry out their evening milking during this peak period after the introduction of smart metering, they may be exposed to increases in electricity costs.

In the United Kingdom, over 4.5 million residential consumers avail of time-of-use (TOU) tariffs. To participate in these programmes, the consumer needs a radio or tele-switch meter connected to the load shifting appliances (Torriti et al., 2010). In July 2007, the Office of Gas and Electricity Metering in the United Kingdom launched the energy demand research project (EDRP). The EDRP consisted of 61 000 residential electricity consumers, of which 18 000 were supplied with smart meters. The use of smart metering and in-home displays resulted in a 3% (0–11% range) overall load reduction and the implementation of TOU tariffs promoted an increase in load shifting of up to 10% (Ofgem, 2011).

Estonia, Finland, France, Ireland, Italy, Malta, the Netherlands, Norway, Portugal, Spain, Sweden and the United Kingdom are all classified as 'Dynamic Movers' in relation to the implementation of smart grid infrastructure. These countries have a clear path towards a full rollout of smart metering. Either the mandatory rollout is already decided or there are major pilot projects that are paving the way for a subsequent decision (Hierzinger et al., 2012). Countries such as Australia and New Zealand have recognised smart metering as a method of improving resource use efficiency and have carried out some early-stage feasibility studies and cost-benefit analysis calculations (Energy Fed NZ, 2010; DRET, 2008). Many of these countries have well-established milk production industries that may be able to use smart grid infrastructure to their advantage by observing that quite large electricity cost increases could occur if a farmer chooses the incorrect electricity tariff. In order to use the smart grid infrastructure, further knowledge is required on the effects of different technologies on the farm electricity consumption and costs across various tariff options.

2.3 Modelling energy consumption

Some whole-farm simulation models published to date contain a sub-model to calculate on-farm energy use. For example, DairyWise is an empirical model integrating all major subsystems of a dairy farm into a whole-farm model (Schils et al., 2007), FarmGHG is an empirical model of carbon and nitrogen flows on dairy farms (Olesen et al., 2006), FarmSim is another model of GHG emissions at the farm scale where on-farm energy use is computed (Saletes et al., 2004) and the Moorepark Dairy Systems Model computes energy costs based on historic consumption values (Shalloo et al., 2004). Although these models contain an energy calculation element for computing total on-farm GHG emissions or farm profitability, they are not suitable for assessing the effects of changes to the system. This is because they are either based on historic consumption trends, do not include sufficient technology specific details, do not account for the impact of ambient temperatures on technology performance, or do not place the electricity consumption in the correct hourly time horizon, making analysis of future time-varying electricity tariffs impossible. Moreover, despite the work of previous authors to bring net present value calculations to support capital investment from the horticulture industry (Aramyan et al., 2007) to the dairy farm (Gebrezgabher et al., 2012), the simple payback method is generally used to support investment in new technologies on dairy farms. This method computes a payback figure (in years) by dividing the estimated annual monetary saving generated by an item of energy efficient technology by the purchase costs of that technology. This payback method, however, can be misleading because critical factors such as the value of money over time, increases in electricity price and the interest rate of borrowed capital from financial institutions are not taken into account, all of which can affect the effective return on investment of a technology.

3 Strategies to reduce energy use in dairy farming

The factors described above, that are currently acting on dairy farming businesses, will create an unprecedented level of uncertainty around electricity consumption and costs, and associated GHG emissions on dairy farms. Providing farmers with strategies to reduce electricity consumption, costs and emissions can help turn the policy measures described above into an advantage for dairy farmers.

This case study focuses on two main groups of strategies to reduce electricity consumption in dairy milking facilities from economic and environmental viewpoints. First, the 'cost strategies' consisted of measures that could save on-farm electricity costs but not energy or related emissions, for example, moving to a new electricity tariff. Another 'cost strategy' is decoupling large electricity uses, such as water heating, from milking times and shifting them to off-peak periods when electricity price is lower. This could also be achieved by adjusting milking start times to avoid consuming electricity at peak electricity prices. Second, the 'energy strategies' changed electricity consumption through farm management changes and application of technologies which aimed to reduce electricity consumption, associated costs and GHG emissions. Possible 'energy strategies' are the use of pre-cooling of milk and solar thermal technologies to provide hot water for washing of milking equipment.

3.1 Electricity consumption model

To examine the potential of the above described strategies to reduce electricity consumption and/or costs we used a model for simulating electricity consumption, associated costs and GHG emissions on dairy farms (MECD) (Upton et al., 2014). The MECD is a mechanistic mathematical representation of electricity consumption that simulates farm equipment on an hourly and monthly basis under the following headings: milk cooling system, water heating system, milking machine system, lighting system, water pump system and the winter housing facilities. The main inputs to the model are milk production, cow number and capacity of the milk cooling system, milking machine system, water heating system, lighting systems, water pump systems and the winter housing facilities as well as details of the management of the farm (e.g. season of calving, frequency of milking and milking start time). The energy consumption of each of the 7 infrastructural systems described above was computed using the MECD in a 12 x 24 matrix structure that simulated a representative day for each month of the year (12 months x 24 hours). Electricity tariffs were compiled in an identical 12 x 24 matrix. Dairy farm electricity costs were then calculated by multiplying the energy consumption matrix by the tariff matrix. There are three types of electricity tariff discussed in this case study.

Flat and day and night tariffs

Two commonly used existing tariff structures were used in this analysis (Flat and Day&Night) to act as reference tariffs, and to highlight any potential cost savings or increases that may occur due to the implementation of future tariffs. A Flat tariff implied electricity price of €0.16/kWh throughout the year, whereas a Day&Night tariff implied a price of €0.16/kWh from 09:00 to 00:00 h and of

€0.08/kW from 00:00 to 09:00 h. The mean electricity price on the Day&Night Tariff, therefore, was €0.13/kWh.

Real time pricing (RTP) tariffs

Real-time pricing (RTP) of electricity implies a dynamically varying electricity price from hour to hour, from day to day and from season to season. The price deviations are based on the national grid load and demand. The Single Electricity Market (SEM) is the wholesale electricity market operating in Ireland. As a gross mandatory pool market operating with dual currencies and in multiple jurisdictions the SEM represents the first market of its kind in the world (SEMO, 2013). The system marginal electricity price (SMP) for 2010 was downloaded from the SEMO website and used as a basis for the RTP tariff. Electricity prices in the European Union area have changed very little between 2010 and 2016 (8% increase) (Eurostat, 2016). The SMP, however, does not reflect the price paid by the consumer, as other charges apply, such as transmission costs, balancing costs, distribution costs and retail margin. Costs for these additions in 2010 were sourced from Deane et al. (2013). The RTP tariff, therefore, was computed as

$$RTP(i,j) = SMP(i,j)+Tc+Bc+Dc+Rm \qquad [1]$$

Where $RTP(i,j)$ is the real-time price of electricity in month i (1-12) and hour j (1-24) (€/kWh); $SPM(i,j)$ is the SMP in month i (1-12) and hour j (1-24); Tc is transmission cost, taken as €0.008/kWh; Bc is balancing cost, taken as €0.003/kWh; Dc is distribution cost taken as €0.051/kWh, and Rm is retail margin taken as €0.017. The mean electricity cost of the RTP was €0.13/kWh (range 0.11–0.30 €/kWh). This tariff varied from month to month and from hour to hour due to the dynamic nature of the SMP. Figure 1 shows the RTP variation by month and by hour.

3.2 Model inputs

The electricity consumption and related costs of a small farm (SF) with 45 milking cows, a medium farm (MF) with 88 milking cows and a large farm (LF) with 195 milking cows was simulated using the MECD. Background data from an energy study of these farms presented by Upton et al. (2013) was used to populate the MECD with data pertaining to the infrastructural configuration on each of these three farms. The SF, MF and LF were spring calving herds operating grass-based milk production systems with low supplementary feed input. Milk production was 255 278 L for SF; 499 898 L for MF and 774 089 L for LF. All farms engaged herringbone milking plants with two stalls per milking unit and were fitted with oil lubricated centrifugal vane vacuum pumps without

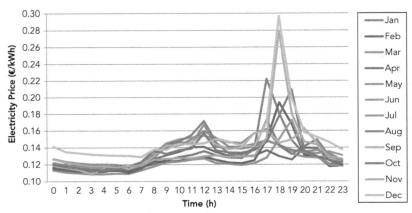

Figure 1 Graph of electricity price (€/kWh) by hour for an average day in each month of the year (2010) for the real-time pricing (RTP) tariff used in this analysis.

variable speed control. Milking parlour size varied from 8 units on SF, 14 units of MF and 24 units on LF.

3.3 Farm profitability module

A profitability module was added to the MECD to assess the return on investment (ROI) arising from a specific investment in an item of energy efficient technology. The profitability module used average financial performance data and variable and fixed cost data from Irish dairy farms. The discounted net profit, tax paid and loan interest paid on borrowed capital were included in the module. ROI is a performance measure of the efficiency of each technology investment scenario. The ROI is calculated by dividing the discounted net profit before tax and interest by the total initial asset value estimated based on fair market value and is displayed as a percentage (McDonald et al., 2013). The ROI is used in this analysis to provide a metric of how effectively each technology investment scenario used capital invested to generate profit. The ROI will be higher for investments requiring lower levels of capital investment to deliver relatively higher equipment running cost savings.

3.4 Energy strategy description

Technologies chosen for analysis in this case study were milk cooling systems and water heating systems. In the analysis presented it was assumed that there was a requirement for investment in a milk cooling system, milking system vacuum pumps and a water heating system. The outputs of the MECD of total

Table 1 Total technology investment costs for the base and three technology investment strategies on three types of farms, SF (small farm), MF (medium farm) and LF (large farm)

Strategy[1]	SF (€)	MF (€)	LF (€)
0-2 Base	20,159	23,977	29,715
4-2 DXPHE	21,243	25,551	32,107
5-2 IBPHE	23,330	28,083	35,586
6-2 Solar	23,733	28,201	35,136

[1] Base = investment in direct expansion (DX) milk cooling system, standard milking system vacuum pumps and electric water heating system; DXPHE = as per Base with the addition of a milk pre-cooling system; IBPHE = Ice Bank (IB) milk cooling system with the addition of a pre-cooling system; Solar = as per Base with the addition of solar thermal panels.

annual electricity consumption and related costs were computed for each technology scenario. The farms morning milking time was set to 07:00 h and the evening milking time was set to 17:00 h. The base level of investment in strategies 4-2, 5-2 and 6-2 (Table 1) included investment in a direct expansion (DX) milk cooling system, standard milking system vacuum pumps and electric water heating system. Table 2 presents data pertaining to three other alternative investment strategies: 4-2 termed DXPHE where the investment included a DX milk cooling system with the addition of a milk pre-cooling system to cool milk to 15°C with well water via a plate heat exchanger; strategy 5-2 termed IBPHE utilised an Ice Bank (IB) milk cooling system with the addition of a pre-cooling system to cool milk to 15°C with well water; and 6-2 termed Solar, which was similar to the Base level of investment with the addition of solar thermal panels for water heating. Reference technology costs were sourced from Irish government reference cost guidelines (DAFM, 2013). Details of these costs are included in Table 1.

4 Results, analysis and recommendations

4.1 Cost reduction strategies

Adjusting milking start time had the effect of moving consumption to off-peak periods. However, milking earlier in the morning and later in the evening (Strategy 1-1 in Table 2) also reduced the simulated annual electricity consumption and related GHG emissions by between 5% and 7%, depending on farm size, relative to the base case (Strategy 0-1 in Table 2). This difference was accounted for by the variation in performance of the milk cooling system, because the efficiency of cooling increased as the ambient temperature decreased. Indeed, the strategy of milking earlier in the morning and later in the evening was effective in reducing electricity costs by 36% for a SF with 45 milking cows, 30% for a MF with 88 milking cows and 34% for a LF with 195 cows. Work has been carried out to describe the effect of RTP electricity tariffs in the residential sector (Tiptipakorn and Wei-Jen, 2007, Allcott, 2011, Ericson, 2011, Finn et al., 2012, Torriti, 2012) and the industrial sector (Avci et al., 2013, Wang and Li, 2013, Finn and Fitzpatrick, 2014). Similar analysis has not been reported in relation to the agricultural sector. This case study showed that the Day&Night electricity tariff minimised annual electricity costs (Strategy 0-1 in Table 2), while a Flat tariff would increase the electricity costs of the SF by 31%, MF by 34% and LF by 16% (Strategy 2-1 in Table 2). Likewise an investigation of an RTP tariff showed that annual electricity costs would increase by 15% for SF, 18% for MF and 3% for LF, see Strategy 3-1 in Table 2. This information will help dairy farmers choose the most appropriate electricity tariff in future. While these analyses were carried out with respect to the Irish dairy production environment, the results pertaining to the impact of various future electricity pricing tariffs and technologies on dairy farm electricity costs are of relevance to dairy industries internationally. This is especially true where dairy farms are expanding and investment in new technology is required. Moreover, these results will be useful in informing decision-making in regions where dairy farms are offered TOU tariffs or dynamic tariffs, as described in the 'Energy efficiency incentives' section of this chapter.

4.2 Energy reduction strategies

An analysis of the 'energy strategies' is presented in Table 2. The strategies DXPHE, IBPHE and solar all reduced electricity consumption, CO_2 emissions and electricity costs. The strategy IBPHE in Table 2 resulted in the largest annual electricity cost saving, 46% for SF, 38% for MF and 45% for the LF; however, the ROI figures were all negative (29% for SF, 23% for MF and 21% for LF). Likewise, even though the solar strategy reduced electricity consumption, CO_2 emissions and electricity costs the ROI figures were all negative, i.e. 225% for the SF, 218% for the MF and 216% for the LF. The most attractive ROI figures resulted from the DXPHE strategy, 17% for SF, 19% for MF and 21% for LF ceteris paribus.

Farmers with about 195 milking cows can reduce their electricity consumption by over 18%, saving over 3 tonnes of CO_2 per annum by using a solar thermal water heater (see strategy 6-2, table 2), however over a ten-year period subsequent to the investment the farmer will have reduced cumulative profitability of 1.8% (€9,200). Likewise, optimising for energy cost on the LF would lead a farmer to choose the IBPHE investment strategy; however, this would lead to a decrease in farm profitability of 0.3% (€1,761) over a period of ten years after investment along with increased electricity consumption and associated emissions. Investment in technologies such as pre-cooling coupled with direct expansion milk cooling systems (as in DXPHE from Table 2), however,

Table 2 Summary of the impact of 'cost' and 'energy' strategies on total annual electricity consumption (kWh) associated CO_2 emissions (kg) and associated costs (€) on three farm sizes SF (small farm), MF (medium farm) and LF (large farm) as reported in this case study. Costs are based on the Day&Night tariff for all strategies except 2-1 and 3-1

Strategy number	Farm name	Strategy name[1]	Energy (kWh)	(% change)	CO_2	(% change)	Costs* (€)	(% change)	ROI (%)
0-1	SF	Base	8,498	0	4,504	0	933	0	–
	MF		20,631	0	10,934	0	2,184	0	–
	LF		32,407	0	17,176	0	3,586	0	–
1-1	SF	Milk at 5 am & 8 pm	−618	−7	−328	−7	−334	−36	–
	MF		−1,006	−5	−533	−5	−660	−30	–
	LF		−1,729	−5	−916	−5	−1,206	−34	–
2-1	SF	Move to Flat tariff	0	0	0	0	427	31	–
	MF		0	0	0	0	1,117	34	–
	LF		0	0	0	0	664	16	–
3-1	SF	Move to RTP tariff	0	0	0	0	166	15	–
	MF		0	0	0	0	468	18	–
	LF		0	0	0	0	125	3	–
0-2	SF	Base[2]	10,413	0	5,519	0	1,445	0	–
	MF		25,252	0	13,384	0	3,334	0	–
	LF		32,670	0	17,315	0	4,571	0	–
4-2	SF	DXPHE	−2,876	−28	−1,524	−28	−570	−39	17
	MF		−5,644	−22	−2,991	−22	−1,083	−32	19
	LF		−9,010	−28	−4,775	−28	−1,714	−37	21
5-2	SF	IBPHE	−2,706	−26	−1,434	−26	−667	−46	−9
	MF		−5,258	−21	−2,787	−21	−1,259	−38	−3
	LF		−8,489	−26	−4,499	−26	−2,044	−45	−1
6-2	SF	Solar	−806	−8	−427	−8	−64	−4	−25
	MF		−2,270	−9	−1,203	−9	−182	−5	−18
	LF		−5,764	−18	−3,055	−18	−461	−10	−16

[1]Strategy 1-1 is additive with 2-1 and 3-1. Strategy 6-2 is additive with 4-2 and 5-2.
[2]Base = investment in direct expansion (DX) milk cooling system, standard milking system vacuum pumps and electric water heating system; DXPHE = as per Base with the addition of a milk pre-cooling system; IBPHE = Ice Bank (IB) milk cooling system with the addition of a pre-cooling system; Solar = as per Base with the addition of solar thermal panels.

serves both purposes. The DXPHE strategy reduced electricity consumption by 28%, increased overall ten-year profitability by 0.8% (€3,960) and reduced annual CO_2 emissions by 4.8 tonnes on LF. These results broadly agree with those of Vellinga et al. (2011), who noted that the use of pre-cooling equipment was a cost-effective method to reduce GHG emissions.

4.3 Future technology requirements for dairy farms

Clearly a balance is required between the ability of a technology to reduce electricity costs and its initial capital investment. One example of this balance is the application of optimal control to pre-cooling systems, in order to reduce electricity consumption and associated costs of the milk cooling system. This technology is inexpensive to implement and can reduce electricity consumption by 34% (Murphy et al., 2013). This example ideally represents the path that technology development for dairy farms should follow. Some dairy farmers are currently unaware of the energy efficiency of their milk harvesting equipment. Therefore, technology development should focus on two key areas: 1) energy efficiency by design, where products are designed with energy efficiency in mind. For example, water heaters should be designed with minimal heat loss, and the capacity of equipment should match the demand, especially in the case of pre-cooling systems on larger farms, 2) improved management of dairy farm technologies for energy efficiency. This must encompass built-in active energy monitoring to help with DSM of electricity use at farm level. For example, if farmers are aware that their pre-cooling systems or water heating systems are operating with reduced efficiency, they will be more likely to take remedial action.

5 Sustainable water use in dairy production

5.1 Types of water use

We reviewed two types of water use: the use of soil moisture due to evapotranspiration (green water) and the use of ground and surface water (blue water) for crop irrigation, or as drinking, processing or cleaning water. The WF concept of Hoekstra et al. (2010) also quantifies grey water use, which is a proxy for the degree of freshwater pollution due to wastewater discharges (grey water) (Hoekstra et al., 2011). Grey water represents an emission which can be better represented in other impact categories in a life cycle assessment, such as eutrophication (Milà i Canals et al., 2009, Pfister et al., 2009, Jefferies, 2012).

5.2 Total water footprint

Mekonnen and Hoekstra (2010) estimated that the production of 1 kg of Irish milk, on average, required 670 L of water, of which 633 L was green water and 37 L was blue water. A study by Murphy et al. (2017) reported a figure of 690 L of water per kg FPCM, of which 684 L was green water and 6 L was blue.

The total blue WF (BWF) was 6.4 litres of water per kg of FPCM in the study of Murphy et al. (2017) and 83% (5.3 L/kg FPCM) of this related to the on-farm BWF which refers to the volume of water used to facilitate the milk harvesting processes and water consumed by livestock within the farm gate.

A Dutch study of milk production from a model farm with intensive irrigation of on-farm grass and maize land, not representative of Dutch farming systems, reported that the production of one kg FPCM required 66.4 L of consumptive blue water (De Boer et al., 2013), of which 55% (36.8L/kg FPCM) was for irrigated on-farm grass growth, 18% (12.1 L/kg FPCM) was for irrigated on-farm maize production, 16% (10.3 L/kg FPCM) was for production of concentrates and 8% (5.4 L/kg FPCM) was for dairy husbandry.

Ridoutt et al. (2010) reported a blue WF of 14.1 L per L of milk produced in a temperate area of Australia on a pasture-based system supplemented by purchased hay and grain. Eighty three per cent (11.7 L/kg) of this blue WF related to on-farm water use (i.e. drinking water and milking processes).

A New Zealand study (Zonderland-Thomassen and Ledgard, 2012) compared the WF of dairy farming in two contrasting regions: Waikato (North Island, non-irrigated, moderate rainfall) and Canterbury (South Island, irrigated, low rainfall). The total volumetric WF was 945 L and 1084 L/kg FPCM for the two regions, respectively. The Waikato dairy system had a greater green WF, 72%, whereas in the Canterbury farm system green water constituted 46% of the WF. For both regions it was demonstrated that the green WF was associated with feed sources and grass growth. The blue water contribution in the Canterbury region was associated with irrigated pasture.

A study carried out by Bord Bia and Cranfield University found that between 7.8-8.6 L of blue water per litre of FPCM and 584–635 L of green water per litre of FPCM was required for a spring calving system in Ireland (Hess et al., 2012). In both the study of Hess et al. (2012) and the study by Murphy et al. (2017), there was a relatively large green water component (about 99% green) due to the rain-fed grass-based system of farming in Ireland which required fewer inputs by way of concentrates or other forages, thus there was a lesser demand for concentrates from irrigated crops. As well as this, only a small proportion of the components required for the production of concentrate required irrigation.

The difference in results between the studies described above can be explained by differences in the methods of assessing a WF, terminologies used and in the resolution of data used. The uncertainties in the different methods and models are a general problem in livestock-water assessments. Although hydro-logical models can generate the necessary data to estimate the consumptive water use (CWU), they often operate at a higher spatial resolution than is required to provide insights into local environmental impacts following changes in CWU (Ran et al., 2016).

6 Conclusions: the relevance of energy reduction and water management strategies to dairy farm sustainability

It is vital for the sustainability of our food-production systems that the abstractions of water for agricultural production (i.e. blue water use) and fossil energy consumptions are quantified, optimised and managed responsibly, in order to avoid local or seasonal water shortages, nationwide water stresses and fossil energy depletion.

The water stress index (WSI) is used to assess the related impact of blue water use; it is considered a mid-point indicator assessing water deprivation and applies to blue water only (Pfister et al., 2009). Blue water describes water abstracted from an underground source, freshwater body or reservoir for irrigation, animal husbandry or cleaning, whereas green water refers to rainwater or water stored in soil that is consumed in the growing of plants used to feed livestock. A WSI indicates the water consumption impacts in relation to water scarcity. The index stems from the water withdrawal to water availability (WTA) ratio. WTA is defined as the ratio of the total annual freshwater withdrawal for human uses in a specific region to the annually available renewable water supply in that region (Frischknecht et al., 2006), and varies across seasons. The method can be applied at the country, region or watershed level. All total volumes of blue water in each region are multiplied by their specific regional WSI to calculate a global average WSI. In order to calculate the stress weighted WF, each source of blue water use is multiplied by the relevant WSI and summed across the supply chain of the food-production system. To assess the global impact of freshwater use, the weighted WF is normalised by dividing it by the global average WSI, giving a quantitative comparison of the pressure exerted from freshwater use through the consumption of a product, relative to the impact of consuming 1 kg of water across the globe (Ridoutt et al., 2010). The severity of water scarcity of water sheds is ranked as follows: WSI , 0.1 low; 0.1 0.1 # WSI , 0.5 moderate; 0.5 # WSI , 0.9 severe and WSI . 0.9 extreme (Pfister et al., 2009). Hence, reduction of the WSI adjusted footprint of a product encourages reduction of blue water use in favour of green water use. This can be accomplished, in the case of milk production, by sourcing animal feeds with low levels of irrigation and through on-farm water reductions through water conscious management practices and elimination of water leaks. It has been estimated that approximately 26% of the stock drinking water on pasture-based dairy farms is wasted through leakage (Higham et al., 2017). The demands for on-farm water can vary seasonally; times of peak milk production can coincide with periods of low water supply in the summer months due to less precipitation and groundwater recharge. During these periods the consumption of freshwater can be lessened through water use monitoring together with improved milk harvesting technologies (i.e. water recycling) (Murphy et al., 2014).

In this chapter we have examined the effect of on-farm energy reduction strategies. By combining strategies 4-2 and 6-2 (Table 2) a saving of over 45% in annual electricity consumption and related GHG emissions could be achieved, thereby contributing to dairy farm energy efficiency, GHG emission reductions and result in an increase in renewable energy production.

The electricity-related GHG emissions from the average dairy farm studied by Upton et al. (2013) was found to be about 23.7 tonnes CO_2 equivalent (CO_2-eq) per annum. To put this in context, the total emissions from a seasonal grass–based dairy farm were found to be 11.7 tonnes of CO_2-eq per tonne of milk solids (O'Brien et al., 2012). This would result in total annual emissions of about 515 tonnes CO_2-eq for the average farm of the 22 farms studied in Upton et al. (2013), of which only about 5% results from electricity consumption.

The impact of reducing electricity consumption on reducing energy depletion, therefore, is relatively more important than on reduction of GHG emissions because, in dairy production CH_4 and N_2O are the dominant GHGs (Basset-Mens et al., 2009; Cederberg and Flysjö, 2004; O'Brien et al., 2012; Thomassen et al., 2009; Wells, 2001). In this analysis, the indicators of GHG emissions and energy consumption per litre of milk were used as proxies for the environmental performance of dairy farms. Electricity consumption is also associated with other environmental impacts (i.e. acidification potential due to SO_2 and NO_x emissions); however, these impacts are related to electricity consumption via a fixed multiplication factor of 2.76 g/kWh for SO_2 and 1.2 g/kWh for NO_x (Huijbregts, 1999).

7 Where to look for further information

Further information on energy and water use efficiency in agriculture can be found in the following links:

- https://ec.europa.eu/energy/en/topics/energy-efficiency/energy-efficiency-directive
- http://ec.europa.eu/environment/water/water-framework/index_en.html
- http://www.fao.org/nr/water/docs/FAO_nexus_concept.pdf
- http://www.unwater.org/topics/water-food-and-energy-nexus/en/
- http://www.fao.org/docrep/014/i2454e/i2454e00.pdf
- http://www.venturesouthland.co.nz/Portals/0/Documents/R_13_3_3227.PDF
- http://www.uwex.edu/energy/dairy.html

8 References

Allcott, H. 2011. Rethinking real-time electricity pricing. *Resource and Energy Economics* 33(4):820–842.

Aramyan, L. H., A. G. J. M. O. Lansink and J. A. A. M. Verstegen. 2007. Factors underlying the investment decision in energy-saving systems in Dutch horticulture. *Agricultural Systems* 94(2):520–7.

Avci, M., M. Erkoc, A. Rahmani and S. Asfour. 2013. Model predictive HVAC load control in buildings using real-time electricity pricing. *Energy and Buildings* 60(0):199–209.

Basset-Mens, C., S. Ledgard and A. Carran. 2005. *First Life Cycle Assessment of Milk Production Systems for New Zealand Dairy Farm Systems.* Paper presented to the Australia New Zealand Society for Ecological Economics Conference, Palmeston North, Massey University, pp. 258–65.

Basset-Mens, C., S. Ledgard and M. Boyes. 2009. Eco-efficiency of intensification scenarios for milk production in New Zealand. *Ecological Economics* 68(6):1615–25.

Busscher, W. 2012. Spending our water and soils for food security. *Journal of Soil and Water Conservation* 67(3):228–34.

Calcante, A., F. M. Tangorra, and R. Oberti. 2016. Analysis of electric energy consumption of automatic milking systems in different configurations and operative conditions. *Journal of Dairy Science* 99(5):4043–7.

Cederberg, C., and A. Flysjö. 2004. Life cycle inventory of 23 dairy farms in southwestern Sweden. SIK Report Nr 728. The Swedisch Institute for Food and Biotechnology, Göteborg, Sweden.

DAFM. 2013. Department of Agriculture Food and the Marine, National Reference Costs, Dublin, Ireland.

Deane, P., J. Fitzgerald, L. Malaguzzi-Valeri, A. Touhy, and D. Walsh. 2013. Working Paper No. 452. Irish and British historical electricity prices and implications for the future. Economic and Social Research Institute (ESRI), Dublin, Ireland.

De Boer, I. J. M., I. E. Hoving, T. V. Vellinga, G. W. J. Van de Ven, P. A. Leffelaar, and P. J. Gerber. 2013. Assessing environmental impacts associated with freshwater consumption along the life cycle of animal products: the case of Dutch milk production in Noord-Brabant. *The International Journal of Life Cycle Assessment* 18(1):193–203.

De Vries, M., and I. J. M. De Boer. 2010. Comparing environmental impacts for livestock products: A review of life cycle assessments. *Livestock Science* 128(1–3):1–11.

DRET. 2008. Cost-benefit analysis of options for a national smart meter roll-out: Phase Two—Regional and detailed analyses regulatory impact statement. Department of Resources Energy and Tourism (DRET), Canberra, Australia.

EC. 2008. Communication from the Commission to the European Parliament, the Council, the European Economic and Social Committee and the Committee of the Regions - 20 20 by 2020 - Europe's climate change opportunity, European Commission, Brussels, Belgium.

Energy Federation of New Zealand. 2010. Developing our energy potential, New Zealand Energy Efficiency and Conservation Strategy. Energy Federation of New Zealand, Wellington, New Zealand.

Ericson, T. 2011. Households' self-selection of dynamic electricity tariffs. *Applied Energy* 88(7):2541–7.

EU. 2006. Directive 2006/32/EC of the European Parliment and of the council on energy end-use efficiency and energy services and repealing Council Directive 93/76/EEC. Official Journal of the European Union.

Eurostat. 2016. Eurostat energy price tables, accessed 27-09-2016, http://ec.europa.eu/eurostat/web/energy/data/database

FAO. 2009. The state of food and agriculture. Livestock in balance. Sales and marketing group, communication division, FAO, Rome, Italy.

Finn, P., and C. Fitzpatrick. 2014. Demand side management of industrial electricity consumption: Promoting the use of renewable energy through real-time pricing. *Applied Energy* 113(0):11-21.

Finn, P., C. Fitzpatrick, and D. Connolly. 2012. Demand side management of electric car charging: Benefits for consumer and grid. *Energy* 42(1):358-63.

Frischknecht, R., R. Steiner, A. Braunschweig, N. Egli, and G. Hildesheimer. 2006. Swiss ecological scarcity method: the new version 2006. Berne, Switzerland.

Gebrezgabher, S. A., M. P. M. Meuwissen, and A. G. J. M. Oude Lansink. 2012. Energy-neutral dairy chain in the Netherlands: An economic feasibility analysis. *Biomass and Bioenergy* 36(0):60-8.

Hartman, K., and R. Sims. 2006. Saving energy on the dairy farm makes good sense. Proceedings of the 4th Dairy3 Conference held at Hamilton New Zealand. Centre for Professional Development and Conferences, Massey University, Palmerston North, New Zealand, pp. 11-22.

Hess, T. M., J. Chatterton, and A. Williams. 2012. The Water Footprint of Irish Meat and Dairy Products. Bord Bia. https://dspace.lib.cranfield.ac.uk/bitstream/1826/8756/3/The_Water_Footprint_of_Irish_Meat_and_Dairy_Products-2012.pdf

Hierzinger, R., M. Albu, H. van Elburg, A. J. Scott, A. Lazicki, L. Penttinen, F. Puente, and H. Sale. 2012. European Smart Metering Landscape Report, SmartRegions Deliverable 2.1. Accessed24 May 2013. http://www.energyagency.at/fileadmin/dam/pdf/projekte/klimapolitik/SmartRegionsLandscapeReport2012.pdf

Higham, C., D. Horne, R. Singh, B. Kuhn-Sherlock, and M. Scarsbrook. Water use on non-irrigated pasture based dairy farms *Journal of Dairy Science*. 100(1):828-840.

Hoekstra, A. Y., Chapagain, A. K., Aldaya, M. M., and Mekonnen, M. M. 2011. The Water Footprint Assessment Manual. Setting the Global Standard.

Houston, C., S. Gyamfi, and J. Whale. 2014. Evaluation of energy efficiency and renewable energy generation opportunities for small scale dairy farms: A case study in Prince Edward Island, Canada. *Renewable Energy* 67:20-9.

Huijbregts, M. A. J. 1999. Life-cycle impact assessment of acidifying and eutrophying air pollutants. Calculation of characterisation factors with RAINSLCA. Interfaculty Department of Environmental Science, Faculty of Environmental Science, University of Amsterdam.

Jefferies, D., I.Munoz, Hodges, J., V. J. King, M. Aldaya, A. E. Ercin, L. M. I. Canals and A. Y. Hoekstra. 2012. Water Footprint and Life Cycle Assessment as approaches to assess potential impacts of products on water consumption. Key learning points from pilot studies on tea and margarine. *Journal of Cleaner Production* 33:155-66.

Ludington, D., and E. L. Johnson. 2003. Dairy farm energy audit summary. New York State Energy Research and Development Authority https://www.nyserda.ny.gov/-/media/Files/.../Energy-Audit.../dairy-farm-energy.pdf

McDonald, R., L. Shalloo, K. M. Pierce, and B. Horan. 2013. Evaluating expansion strategies for startup European Union dairy farm businesses. *Journal of Dairy Science* 96(6):4059–69.

Mekonnen, M., and A. Hoekstra. 2010. The green, blue and grey water footprint of farm animals and animal products. http://waterfootprint.org/media/downloads/Report-48-WaterFootprint-AnimalProducts-Vol2.pdf

Milà i Canals, L., J. Chenoweth, A. Chapagain, S. Orr, A. Anton, and R. Clift. 2009. Assessing freshwater use impacts in LCA: Part I-inventory modelling and characterisation factors for the main impact pathways. *The International Journal of Life Cycle Assessment* 14(1):28–42.

Molden, D., M. Vithanage, C. de Fraiture, J. M. Faures, L. Gordon, F. Molle, and D. Peden. 2011. 4.21 - Water Availability and Its Use in Agriculture. In Treatise on Water Science. W. Editor-in-Chief: Peter, ed. Elsevier, Oxford, pp. 707–32.

Morison, K., W. Gregory, and R. Hooper. 2007. Improving Dairy Shed Energy Efficiency, Technical Report. New Zealand Centre for Advanced Engineering (CAENZ). Christchurch, New Zealand.

Murphy, E., I. J. M. de Boer, C. E. van Middelaar, N. M. Holden, L. Shalloo, T. P. Curran, and J. Upton. 2017. Water footprinting of dairy farming in Ireland. *Journal of Cleaner Production* 140, Part 2:547–555.

Murphy, E., J. Upton, N. M. Holden, and T. P. Curran. 2014. Direct water use on Irish dairy farms. In Biosystems Engineering Research Review 19. University Collage Dublin, Ireland.

Murphy, M. D., J. Upton, and M. J. O'Mahony. 2013. Rapid milk cooling control with varying water and energy consumption. *Biosystems Engineering* 116(1):15–22.

O'Brien, D., L. Shalloo, J. Patton, F. Buckley, C. Grainger, and M. Wallace. 2012. A life cycle assessment of seasonal grass-based and confinement dairy farms. *Agricultural Systems* 107:33–46.

Oenema, J., H. van Keulen, R. L. M. Schils, and H. F. M. Aarts. 2011. Participatory farm management adaptations to reduce environmental impact on commercial pilot dairy farms in the Netherlands. *NJAS - Wageningen Journal of Life Sciences* 58(1–2):39–48.

Ofgem. 2011. Energy Demand Research Project: Final Analysis. Office of gas and electricity metering, Report no. 60163857. Hertfordshire, United Kingdom.

Olesen, J. E., K. Schelde, A. Weiske, M. R. Weisbjerg, W. A. H. Asman, and J. Djurhuus. 2006. Modelling greenhouse gas emissions from European conventional and organic dairy farms. *Agriculture, Ecosystems and Environment* 112(2–3):207–20.

Pfister, S., A. Koehler, and S. Hellweg. 2009. Assessing the Environmental Impacts of Freshwater Consumption in LCA. *Environmental Science & Technology* 43(11):4098–104.

Postel, S. L. 2000. Entering an era of water scarcity: The challenges ahead. *Ecological Applications* 10(4):941–8.

Rae, A. N. 1998. The effects of expenditure growth and urbanisation on food consumption in East Asia: a note on animal products. *Agricultural Economics* 18(3):291–9.

Ran, Y., M. Lannerstad, M. Herrero, C. E. Van Middelaar, and I. J. M. De Boer. 2016. Assessing water resource use in livestock production: A review of methods. *Livestock Science* 187:68–79.

Ridoutt, B. G., S. R. O. Williams, S. Baud, S. Fraval, and N. Marks. 2010. Short communication: The water footprint of dairy products: Case study involving skim milk powder. *Journal of Dairy Science* 93(11):5114-17.

Rockström, J., L. Karlberg, S. P. Wani, J. Barron, N. Hatibu, T. Oweis, A. Bruggeman, J. Farahani, and Z. Qiang. 2010. Managing water in rainfed agriculture—The need for a paradigm shift. *Agricultural Water Management* 97(4):543-50.

Rosegrant, M. W., X. Cai, and S. A. Cline. 2002. World water and food to 2025: Dealing with scarcity. Intl Food Policy Res Inst.

Saletes, S., J. Fiorelli, N. Vuichard, J. Cambou, J. E. Olesen, S. Hacala, M. Sutton, J. Fuhrer, and J. F. Soussana. 2004. Greenhouse gas balance of cattle breeding farms and assessment of mitigation options. In: Kaltschmitt, M., Weiske, A. (eds), *Greenhouse Gas Emissions from Agriculture*. Mitigation Options and Strategies. Institute for Energy and Environment, Leipzig, pp. 203-8.

Schils, R. L. M., M. H. A. de Haan, J. G. A. Hemmer, A. van den Pol-van Dasselaar, J. A. de Boer, A. G. Evers, G. Holshof, J. C. van Middelkoop, and R. L. G. Zom. 2007. DairyWise, A Whole-Farm Dairy Model. *Journal of Dairy Science* 90(11):5334-46.

SEMO. 2013. Single Electricity Market Operator website. Accessed 20 Dec 2013. http://www.sem-o.com/AboutSEMO/Pages/default.aspx

Shalloo, L., P. Dillon, M. Rath, and M. Wallace. 2004. Description and validation of the Moorepark Dairy System Model. *Journal of Dairy Science* 87(6):1945-59.

Smith, P., D. Martino, Z. Cai, D. Gwary, H. Janzen, P. Kumar, B. McCarl, S. Ogle, F. O'Mara, C. Rice, B. Scholes, O. Sirotenko, M. Howden, T. McAllister, G. Pan, V. Romanenkov, U. Schneider, S. Towprayoon, M. Wattenbach, and J. Smith. 2008. Greenhouse gas mitigation in agriculture. *Philosophical Transactions of the Royal Society B: Biological Sciences* 363(1492):789-813.

Thomassen, M. A., M. A. Dolman, K. J. van Calker, and I. J. M. de Boer. 2009. Relating life cycle assessment indicators to gross value added for Dutch dairy farms. *Ecological Economics* 68(8-9):2278-84.

Tiptipakorn, S., and L. Wei-Jen. 2007. A Residential Consumer-Centered Load Control Strategy in Real-Time Electricity Pricing Environment. In Proc. Power Symposium, 2007. NAPS '07. 39th North American, pp. 505-10.

Torriti, J., M. G. Hassan, and M. Leach. 2010. Demand response experience in Europe: Policies, programmes and implementation. *Energy* 35(4):1575-83.

Torriti, J. 2012. Price-based demand side management: Assessing the impacts of time-of-use tariffs on residential electricity demand and peak shifting in Northern Italy. *Energy* 44(1):576-83.

UNEP. 2007. Global Environment Outlook - Geo 4: Environment for Development. United Nations Environment Programme, Valletta, Malta.

Upton, J., J. Humphreys, P. W. G. G. Koerkamp, P. French, P. Dillon, and I. J. M. D. Boer. 2013. Energy demand on dairy farms in Ireland. *Journal of Dairy Science* 96(10):6489-98.

Upton, J., M. Murphy, L. Shalloo, P. W. G. Groot Koerkamp, and I. J. M. De Boer. 2014. A mechanistic model for electricity consumption on dairy farms: Definition, validation, and demonstration. *Journal of Dairy Science* 97(8):4973-84.

Vellinga, T. V., M. H. A. de Haan, R. L. M. Schils, A. Evers, and A. van den Pol-van Dasselaar. 2011. Implementation of GHG mitigation on intensive dairy farms: Farmers' preferences and variation in cost effectiveness. *Livestock Science* 137(1-3):185-95.

Wang, Y., and L. Li. 2013. Time-of-use based electricity demand response for sustainable manufacturing systems. *Energy* 63(0):233–44.

Wells, C. 2001. Total energy indicators of agricultural sustainability: Dairy farming case study. Technical Paper 2001/3 prepared for the New Zealand Ministry of Agriculture and Forestry, Wellington, New Zealand.

Zonderland-Thomassen, M. A., and S. F. Ledgard. 2012. Water footprinting - A comparison of methods using New Zealand dairy farming as a case study. *Agricultural Systems* 110:30–40.

Chapter 3

Nutritional factors affecting greenhouse gas production from ruminants: implications for enteric and manure emissions

Stephanie A. Terry, Agriculture and Agri-Food Canada, Canada and University of Sydney, Australia; Carlos M. Romero, Agriculture and Agri-Food Canada and University of Lethbridge, Canada; and Alex V. Chaves and Tim A. McAllister, Agriculture and Agri-Food Canada, Canada

1 Introduction

Animal agriculture has been identified as one of the major sources of greenhouse gases (GHGs), accounting for approximately 40% of total agricultural-related emissions (IPCC, 2006) (Fig. 1). Animal production and manure management comprises 26.8% and 31.0%, respectively, of the 7.1 Gt of CO_2-eq that the livestock sector is estimated to produce annually (Gerber et al., 2013). The two central GHGs emitted directly from animal agriculture include methane (CH_4) and nitrous oxide (N_2O) which have 28 and 298 times the global warming potential of CO_2, respectively (Gerber et al., 2013). Livestock CH_4 and N_2O emissions have been estimated to contribute 40% and 48% of livestock sector emissions and ruminants account for 80% of the total livestock sector's emissions (Opio et al., 2013). Enteric fermentation and manure-related CH_4 contribute 82% and 18% of CH_4 related to livestock production, respectively. The main sources of N_2O emissions arise from chemical fertilisers, applied

http://dx.doi.org/10.19103/AS.2020.0067.16

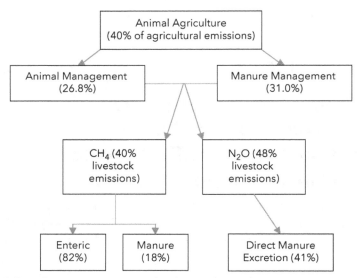

Figure 1 Greenhouse gases associated with animal agriculture.

manure and N deposition from housed animals and manure storage (Adler et al., 2015).

1.1 Greenhouse gas production

Ruminants produce CH_4 as a natural by-product of microbial fermentation, with the biochemical pathways being well documented (Huws et al., 2018). Starches, cell wall polymers and proteins are fermented by a consortium of rumen microbiota into simple sugars and carbon skeletons. Both primary and secondary fermenters convert these products, under anaerobic conditions, into volatile fatty acids (VFA), CO_2 and metabolic hydrogen [H]. Rumen ciliates and anaerobic fungi are two groups of eukaryotes which produce large volumes of [H] and share a commensal relationship with archaea (Guyader et al., 2014). Both protozoa and fungi contain hydrogenosomes which are specialised organelles that are responsible for the conversion of the intermediates of monosaccharide fermentation into [H]. Methanogens play an important role in maintaining a low partial pressure of [H], favouring hydrogenase activity within hydrogenosomes. During fermentation, the reduced co-factors NADH, NADPH and FADH are oxidised and the released [H] is transferred to methanogenic archaea through a series of biochemical pathways to reduce CO_2 to CH_4 (Ungerfeld, 2015b).

Manure CH_4, like enteric CH_4, is produced during anaerobic decomposition of organic matter (OM). Manure is also a significant emitter of N_2O, ammonia

Published by Burleigh Dodds Science Publishing Limited, 2020.

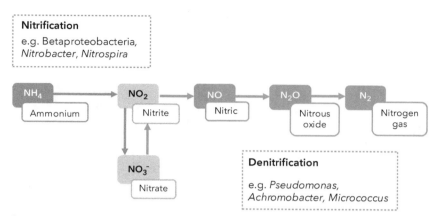

Figure 2 Process of nitrification and denitrification via nitrite pathway. N_2O is a greenhouse gas with a global warming potential that is 298 times that of CO_2.

(NH_3) and nitrous oxides (NO_x). These gases may act as direct or indirect sources of GHGs and environmental pollutants. Factors which influence the concentration of N_2O, NH_3 and NO_x include (i) the type of feed, (ii) manure nutrient profile and (iii) the handling and storage of manure. The conversion of N into gases occurs through the simultaneous nitrification and denitrification process (Fig. 2). Nitrification occurs by both NH_3 oxidising bacteria (i.e. Betaproteobacteria or *Thaumarchaeota* NH_3-oxidizing archaea) and nitrite (NO_2^-) oxidising bacteria (i.e. Alphaproteobacteria *Nitrobacter* and *Nitrospira*). Denitrifying bacteria are phylogenetically diverse and have specific genes coding for their catalytic enzymes (Maeda et al., 2011).

Although not a direct source of GHG, NH_3 emissions from manure should also be considered when assessing the impact of feed additives on air quality. Ammonia arises from the rapid hydrolysis of urea in urine and can also be a precursor to N_2O. Ammonia is highly volatile and can have serious implications for human health when threshold limits of 25-35 ppm are exceeded (National Research Council, 2008). Additionally, dry or wet NH_3 deposition may contribute to soil acidity and eutrophication of surface water (Hünerberg et al., 2013a). Shifting the excretion of N from urine to faeces may be more environmentally beneficial as faecal N is considered a slow-release form of N that is more likely to be captured by soil flora.

1.2 Balancing enteric methane production and manure emissions

Significant interrelationships exist between enteric CH_4 production and both CH_4 and N_2O manure emissions (Knapp et al., 2014). Balancing net emissions produced directly from the rumen or indirectly from manure is challenging.

DIETARY CHANGES	ENTERIC							MANURE					OVERALL	
	Fermentation	Fibre Digestion	N Digestion	Starch Digestion	pH	CH$_4$ PRODUCTION DAILY	CH$_4$ INTENSITY	Starch	Urine N	Faecal N	CH$_4$ EMISSIONS	N$_2$O EMISSIONS	DECREASE IN GHG PRODUCTION	DECREASE IN GHG INTENSITY
Concentrate/Forage														
Increased Concentrate/Forage	↑	↓	NA	↑	↓	~	↓	↑	NA	NA	↑	↑	~	✓
Acidosis	↑	↓	NA	↑	↓	~	-	~	NA	NA	~	-	~	~
High Forage	↓	↑	NA	↓	↑	↑	↑	↓	NA	NA	↓	↓	~	✗
Nitrogen														
DDGS	~	↑	↑	~	~	↓	↓	↑	↑	↑	↑	↑	✗	✗
Fat														
< 6%	~	-	NA	-	~	↓	↓	-	NA	NA	-	-	✓	✓
> 6%	↓	↓	NA	~	↑	↓	-	↑	NA	NA	↑	~	✓	✗
Inhibitors														
Nitrate	~	-	-	~	-	↓	↓	~	~	~	~	~	✓	✓
3NOP	~	-	-	-	~	↓	↓	-	~	~	-	-	✓	✓
PSC														
Tannins	↓	↓	↓	NA	NA	~	~	~	↓	↑	↓	↓	✓	~

Figure 3 Consequences of dietary manipulation on enteric production and greenhouse gas emissions. Symbols indicate: ↑ = increase, ↓ = decrease, - = no change, NA = not applicable, ~ = variable/unknown.

Another important consideration when evaluating dietary manipulation strategies is accounting for the difference in GHG production and GHG intensity. Although global GHG emissions from ruminants have decreased as a factor of animal product (intensity), the total production of GHG has increased, and will continue to do so as the world's domestic ruminant population is projected to increase from 3.2 to 5.3 billion by 2050 (Turk, 2016).

A standard feedlot diet fed to cattle may result in a higher CH$_4$ production (g/d) and a reduced CH$_4$ intensity (g/kg consumable product) than those grazing on pasture (low CH$_4$ production and high CH$_4$ intensity). However, due to low dietary energy content, pasture-raised ruminants produce manure with half the CH$_4$ yield potential of those raised in feedlots (Koneswaran and Nierenberg, 2008), as pasture-raised animals have much lower starch levels in manure (Hales et al., 2013). Additionally, dietary alterations which result in a shift of fermentation from the rumen to the hindgut may decrease enteric CH$_4$ production, but not change overall net GHG emissions. This concept is also known as pollution swapping, in which an alteration in the production of one GHG results in an upstream or downstream change in the emissions of the same or another GHG (Hristov et al., 2013). Nutritional strategies that alter diet digestibility through increasing dietary fermentable carbohydrates, N and fat content can all result in pollution swapping (Fig. 3).

Published by Burleigh Dodds Science Publishing Limited, 2020.

1.2.1 Diet digestibility and fermentable carbohydrates

Diet digestibility is intrinsically linked to enteric and manure GHG production. The more readily a diet is fermented, the lower the nutrient wastage and GHG emission intensity. Factors such as forage quality, forage-to-concentrate ratio and type of concentrate/forage can all contribute to the microbial efficiency of feed digestion. For example, increasing the quantity of concentrates in the diet can reduce enteric CH_4 production through a greater proportion of easily fermentable carbohydrates. This can shift fermentation from acetate which produces [H], towards propionate which utilises [H] in its synthesis and consequently, decreases the availability of [H] for methanogenesis. Increasing fermentable carbohydrates within the diet can also increase digestibility and passage rate. This can both improve productive performance and decrease the amount of OM excreted in the manure. Less OM in manure reduces the amount of substrate available for decomposition and thus the supply of [H] to methanogens, decreasing manure CH_4. Alternatively, increasing the digestibility of the diet can also increase enteric CH_4 on a g/d basis as more substrates are fermented in the rumen and the production of reducing equivalents increases. For example, lactating dairy cows grazing pasture have shown to have an increase in CH_4 production (g/d) and milk yield (van Wyngaard et al., 2018; Muñoz et al., 2015) when supplemented with increasing concentrates (0–8 kg/d).

Increasing concentrate in the diet can increase the protozoa, *Entodinium* in the rumen, a known non-fibrous carbohydrate degrader. Whereas microbes associated with cellulolytic degradation including *Fibrobacter*, *Polyplastron* and *Ostracodinium* decrease as the level of concentrate increases in the diet (Zhang et al., 2017). The diversity and richness of fungal communities were similar in high-forage versus high-concentrate diets; however, the relative abundance of the fungal phyla *Ascomycota*, *Basidiomycota*, *Cercozoa* and *Chytridiomycota* increased, and *Neocallimastigomycota* decreased with increasing proportions of concentrate (Zhao et al., 2018). Other studies also report that fungal communities have been enriched as the forage proportion of the diet increases (Kumar et al., 2015), reflecting their role in the degradation of complex fibre. Changes in eukaryote abundance and diversity are likely to impact methanogen abundance and diversity as eukaryotes produce the [H] required for methanogenesis.

A high-concentrate diet decreased the overall abundance (richness) of the archaeal population, but did not change the range of microbial species (diversity) (Zhang et al., 2017; Mao et al., 2016). Despite their role in methanogenesis, archaeal communities have been reported to show less variation and diversity as a ruminant adapts from a high-forage to a high-concentrate diet (Henderson et al., 2015; Kumar et al., 2015). This may be due to their low density and their less-diverse metabolic capabilities (Kumar et al., 2015; Henderson et al., 2015). However, *Methanomicrobium* (Methanomicrobiales order) and *Methanomicrococcus* (Methanosarcinales order) are reported to be sensitive to dietary changes with both these taxa being absent in high-grain diets and only detected in forage diets (Friedman et al., 2017a). The absence of *Methanomicrococcus* (Methanosarcinales order) in high-grain diets is likely related to an increase in redox potential associated with a lower rumen pH (Friedman et al., 2017a).

Diets composed of highly fermentable substrates can result in conditions where organic acid production by the microbial population exceeds the buffering capacity of the rumen, leading to a prolonged reduction in rumen pH. Ruminal acidosis is characterised by a reduction in microbial diversity and rumen malfunction including decreased feed intake and feed digestibility. Abundance and diversity of bacteria were decreased in sheep ruminal fluid (Li et al., 2017) and ruminal fluid and faeces in dairy cows (Plaizier et al., 2017) with induced sub-acute ruminal acidosis. High-grain diets usually result in an increase in starch and lactate utilisers as well as propionate producers (*Prevotella, Selenomonas, Megasphaera, Streptococcus*) (Plaizier et al., 2017; Zhu et al., 2018). Specifically, *Prevotella* and *Succinivibrionaceae* dominate the rumen of ruminants-fed high-grain diets (Henderson et al., 2015). In contrast, fibrolytic bacteria including *Butyrivibrio, Ruminococcus* and *Fibrobacter* are vulnerable to low pH and usually decrease in abundance in high-grain diets (Zhu et al., 2018). Whilst other studies report a decrease in fungal diversity (Kumar et al., 2015; Tapio et al., 2017a), Ishaq et al. (2017) found that abundance and diversity of rare fungal taxa was increased with diet-induced sub-acute acidosis, including those associated with lactic acid utilisation (*Pichia* and *Candida*). The abundance of the archaeal population has been shown to not change with increasing concentrate in the diet and decreased ruminal pH (Hook et al., 2011). This suggests that they are resilient to changes in pH and only their functional activity is suppressed under conditions of low pH. This may explain why there is a poor relationship between methanogen abundance and CH_4 production (Firkins and Yu, 2015).

Altering the ability of ruminal microbes to degrade feed can increase nutrient loss in the faeces, increasing the amount of OM available for CH_4 production from manure. Although increasing the concentration of ruminal escape starch in the diet can modulate fermentation and potentially reduce methanogenesis, starch digestion can also be limited (<60%) in the lower digestive tract (Haque, 2018). This results in more starch in the faeces, potentially increasing CH_4 emissions during decomposition of the manure. *Ruminococcaceae* was more abundant in manure of cattle-fed processed grain and forage-fed diets, whereas *Prevotella* dominated in manure from cattle-fed unprocessed grain (Shanks et al., 2011). *Bacteroidetes*, as a reflection of their role in the digestion of complex carbohydrates, increased and *Firmicutes* decreased with increased levels of starch in the faeces. The concentration of starch in faeces from cattle-fed unprocessed grains and processed grains was 98.4% and 66.9% higher, respectively, than in those fed forage, suggesting that inadequate grain processing could increase CH_4 emissions from manure (Shanks et al., 2011).

Improving diet digestibility and thereby the potential of the ruminant to obtain nutrients from feed improves overall growth efficiency. This may lead to an increase or decrease in total GHG emissions depending on the types of gases produced and the balance between enteric and manure emissions. Regardless, if there is an actual decrease in GHG when production from the animal is improved (i.e. less days on feed), GHG emission intensity as a proportion of usable product (i.e. meat, milk, wool, etc.) is reduced (Hristov et al., 2013).

1.2.2 Nitrogen content

Ruminants are a significant contributor to the global N budget. In ruminants, nitrogen cycles through a series of complex biogeochemical interactions involving inorganic- and organic-N in feed, manure and soils (Fig. 4; Robertson and Vitousek, 2009). Plants and animals utilise N throughout this cycle; however, they are limited in their ability to deposit it as a product. For example, the conversion of dietary N into consumable protein (i.e. milk, meat) by ruminants is very low (20-30%) and fertiliser-N recovery by cereal crops seldom exceeds 50% of applied N (Fageria and Baligar, 2005). Excess N from animal agriculture results in the release of a large surplus of reactive N into the environment, mainly via NH_3 and N_2O emissions and/or nitrate (NO_3-N) leaching (Galloway et al., 2004; Powell et al., 2011). Pollution of groundwater by NO_3-N, widespread eutrophication and global warming through N_2O emissions

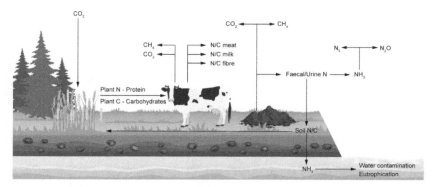

Figure 4 Nitrogen and carbon cycle within ruminant agriculture.

Published by Burleigh Dodds Science Publishing Limited, 2020.

are some well-documented contributions of agricultural-anthropogenic N (Erisman et al., 2013).

The cycle of N within ruminant systems is mainly defined by the transformation of feed-N into milk or meat products, with the remaining N excreted in urine or manure. The concentration of N in urine and faeces depends on the crude protein (CP) content of the diet (Dijkstra et al., 2013). Feeding ruminants to the level of metabolisable protein requirements ensures the best utilisation and the least loss of nutrients (Broderick, 2003). The amount and type of N fed in ruminant diets also has several implications for how it is utilised and excreted by ruminants. Dietary protein supplies both rumen-degradable and -undegradable protein. Rumen-degradable protein is composed of true protein and non-protein N, which when broken down can be utilised for microbial protein synthesis and growth (Bach et al., 2005). Requirements for dietary protein and energy are intrinsically linked, as high-energy diets will stimulate microbial synthesis, enhancing the need for rumen-degradable protein (Broderick, 2003). Whilst changing the CP content in the diet has no obvious direct effect on enteric CH_4, its replacement by carbohydrates can influence emissions.

Replacing protein supply with fermentable carbohydrates is an effective way to reduce urinary N excretion, increase microbial N capture and decrease NH_3 production (Dijkstra et al., 2013). However, replacing dietary CP with fermentable carbohydrates can exacerbate enteric CH_4 production (Sauvant et al., 2011) as increased substrates are available for ruminal methanogenesis. Dijkstra et al. (2013) estimated that higher enteric CH_4 fluxes from increased carbohydrates in the diet are frequently offset by a decrease in N_2O emissions from manure.

Prevotella is a predominant genus within ruminants around the globe (Henderson et al., 2015) and participates in both carbohydrate and N metabolism. Specifically, *Prevotella ruminicola* strain 23 can efficiently degrade hemicellulose and pectin, utilising both NH_3 and peptides as a N source for growth as opposed to amino acids (Kim et al., 2017). Their importance in N metabolism was supported by an observed decline in *Prevotella bryantii* in dairy cattle fed a low-protein diet (Belanche et al., 2012). In this study, the relative abundance of *Ruminococcus albus*, *Ruminococcus flavefaciens*, *Fibrobacter succinogenes* and *Butyrivibrio fibrisolvens* all declined, potentially highlighting the vulnerability of cellulolytic bacteria to N shortages. In contrast, non-cellulolytic bacteria including *Prevotella ruminicola*, *Selenomonas ruminantium*, *Streptococcus bovis*, *Megasphaera elsdenii* and *Aliiglaciecola lipolytica* were not affected by a N shortage which may reflect their low NH_3 requirements for growth (Belanche et al., 2012). Niu et al. (2016) evaluated the effect of two different concentrations of CP (15.2% vs 18.5%) in dairy cows and demonstrated that the low-protein diet decreased total tract digestibility of OM, N and starch

Published by Burleigh Dodds Science Publishing Limited, 2020.

compared to the high protein diet (18.5%). Belanche et al. (2012) indicated that ruminal concentrations of protozoa and methanogens declined with the low-protein diet, whereas Niu et al. (2016) observed no differences in CH_4 emissions as a result of differing protein content in the diet.

Degradation of dietary protein and its assimilation into microbial protein can also contribute to a decrease in available [H] for methanogenesis as this process can both utilise and produce reducing equivalents (Knapp et al., 2014). For example, the synthesis of amino acids can increase with a decrease in methanogenesis, as they act as a [H] sink (Ungerfeld, 2015b). An increase in amino acid production could be related to an increase in the relative abundance of *Bacteroidetes* and *Prevotella* species, both associated with increased proteolytic activity (Martinez-Fernandez et al., 2016). Increasing the proportion of soluble carbohydrates in diets has been related to decreases in branched-chain fatty acids, which are needed for de novo synthesis of amino acids by ruminal microbes. Decreasing availability of branched-chain fatty acids could decrease microbial protein synthesis and microbial growth (Hall and Huntington, 2008) and thereby reduce the extent to which this process acts as an alternative [H] sink to CH_4 production.

Potential interactions between dietary protein content and ruminal methanogenesis remain unclear. Addition of protein to high-fibre diets could increase the efficiency of microbial protein synthesis and reduce the intensity of enteric CH_4 emissions, while increasing N excretion and N_2O missions from manure. Increasing the efficiency of microbial protein synthesis could redirect [H] away from methanogenesis towards the formation of microbial cells and increase the productivity of the ruminants.

1.2.3 Dietary fat

Dietary additives which target microbes involved in methanogenesis may be superior at decreasing enteric CH_4 production without affecting manure emissions. For example, dietary fats may decrease enteric CH_4 production by (i) having a toxic effect on methanogens and protozoa, (ii) replacing fermentable carbohydrates or (iii) providing an alternative [H] sink via biohydrogenation (Beauchemin et al., 2008; Knapp et al., 2014).

The effects of dietary lipids on the rumen are largely dependent on fat composition, concentration and source. Consequently, the impact on the rumen microbial populations varies depending on the nature of the oil. *Methanobrevibacter ruminantium* is the most abundant species of rumen methanogen and was found to be reduced by saturated fatty acids and oleic acid (Henderson et al., 2015; Enjalbert et al., 2017). Addition of linseed and coconut oil decreased CH_4 production, but this was not correlated with changes in the abundance or diversity of the archaeal population (Patra and Yu, 2013; Martin et al., 2016). Similarly, lambs fed linseed oil had a higher relative abundance of *Succinivibrionaceae* (succinate producers) and *Veillonellaceae* (propionate producers) and a decreased abundance of *Ruminococcaceae* (Lyons et al., 2017). An increased abundance of *Succinivibrionaceae* and *Ruminococcaceae* has been associated with low and high CH_4 emitters, respectively (Wallace et al., 2015). *Succinivibrionaceae* produce succinate via utilisation of [H], whereas *Ruminococcaceae* are known hydrogen producers (Wallace et al., 2015). Lambs supplemented with linseed oil had a 19.5% decrease in the relative abundance of *Methanobrevibacter* and a 34.7% increase in *Methanosphaera*, although CH_4 emissions were not measured in this study.

The growth of the rumen fungus, *Neocallimastix frontalis*, was impeded by soybean oil (Boots et al., 2013) and others have found that the fibrolytic bacteria, *Fibrobacter* and *Ruminococcus* are also inhibited by lipids (Enjalbert et al., 2017). Increasing the degree of unsaturated fatty acids in the diet may correlate with decreased protozoal counts (Oldick and Firkins, 2000), and as a result of disrupting the close relationship between protozoa and archaea, change the diversity of archaeal communities (Hristov et al., 2012).

Alteration of the ruminal microbiome due to the type of fat may prompt a high variation in physiological responses including an inhibition of fibre digestion which may decrease CH_4 production. It is well documented that increasing fat content in ruminant diet above 6-7% dry matter (DM) can reduce the digestibility of fibre (Johnson and Johnson, 1995). In continuous culture, an oil (Tucumã) high in oleic acid inhibited CH_4 at 1% (v/v) through suppression of *Fibrobacter* with no alteration in methanogens (Ramos et al., 2018). Decreasing fibre digestibility can result in an increase in manure C, providing substrate for CH_4 emissions. Although, Gautam et al. (2016) found that varying sources of dried distillers grain plus solubles (DDGS) with corn oil (dietary fat of 3-5.5% dietary DM) had no effect on nutrient composition or GHG emissions from manure.

Dietary fat supplementation is an effective enteric CH_4 mitigation strategy. Depending on their fatty acid composition, dietary fats may decrease both daily and the intensity of CH_4 production. However, the main constraint with dietary fat is that the amount that can be supplemented without inhibiting fibre digestion is restricted to ≈6% of dietary DM, so the CH_4 mitigation potential is limited to 10-15%.

2 Case study: Dried distillers' grains plus solubles (DDGS)

A primary example of the importance of a holistic approach to investigating the impact of feeding a dietary additive is shown by research evaluating DDGS. DDGS is a by-product of ethanol production capable of replacing cereal grains in ruminant diets due to its high energy and fat content. Corn

(*Zea mays* L.) and wheat distillers' grains have been shown to decrease CH_4 production as a result of their high oil content. Depending on market demands, DDGS may at times be more economical in least-cost diets than cereal grains.

A meta-analysis by Griffin et al. (2012) found that when DDGS replaced forage, it increased the average daily gain and final body weight of backgrounding cattle. Triticale DDGS could be substituted for barley (*Hordeum vulgare* L.) silage and barley grain in finishing diets with no effect on growth performance or carcass quality in finishing beef steers.

Due to its high fat content, DGGS can successfully decrease enteric CH_4 emissions. This was verified when 35% of barley grain (DM basis) was replaced with corn DDGS and enteric CH_4 production was decreased by 16.4% (% DM intake) in beef cattle fed a barley silage-based diet (McGinn et al., 2009). Similarly, replacing 35% of the barley grain and 5% of the canola meal (*Brassica napus* L.) in a diet with corn DDGS decreased CH_4 production by 15.1% (% DM intake; Hünerberg et al., 2013a). Methane emissions on an intake basis were also decreased (18.0%) in cattle fed a finishing high-grain diet with corn DDGS replacing 40% of barley grain (Hünerberg et al., 2013b). Despite the observed reduction of CH_4 production, these studies reported that including DDGS dramatically increased both N intake and excretion. Additionally, corn DDGS has been shown to decrease the digestibility of starch (Castillo-Lopez et al., 2014), increasing its availability for CH_4 production from manure.

Fibrobacteres and *Bacteroidetes* have been shown to decrease in abundance with inclusion of DDGS in the diet (Castillo-Lopez et al., 2017; Castillo-Lopez et al., 2014). Castillo-Lopez et al. (2017) reported that *Bacteroidetes* were decreased and *Tenericutes* were increased by 20% DDGS (diet DM). Interestingly, the abundance of *Ruminococcaceae* increased with DDGS (Castillo-Lopez et al., 2017), even though it has been suggested that this family is associated with increased CH_4 production (Wallace et al., 2015). Including DDGS at 50% (DM basis) decreased *Succinivibrio* and increased *Bacteroides* and *Prevotella* by −75.4%, 61.0% and 34.6%, respectively.

Though originally seen as an effective GHG mitigation strategy, using a life cycle approach revealed that feeding corn or wheat-based DDGS resulted in a 6.2% and 9.3% increase, respectively, in total GHG emissions compared to a control barley-based diet (Hünerberg et al., 2014). Increased N_2O emissions from manure were found to be a leading factor contributing to total GHG emissions even though manure-related CH_4 emissions were reduced. The increase in N_2O emissions was a result of the increased CP content of the diet and greater N excretion in the urine and the faeces.

For DDGS to be suitable as a GHG mitigation strategy, it needs to be fed at a level that does not exceed the protein requirement of ruminants. However, lower amounts of DDGS in the diet are unlikely to increase fat concentrations

Published by Burleigh Dodds Science Publishing Limited, 2020.

to a level that is sufficient to lower enteric CH_4 emissions (Castillo-Lopez et al., 2017; Judy et al., 2016; Hales et al., 2013).

3 Nitro-based compounds

Nitro-based compounds have been examined for their enteric CH_4 mitigation potential in ruminants. These compounds reduce CH_4 via action as an alternative electron sink or possibly as a direct inhibitor of methanogens. Two dietary feed additives which have received attention for their ability to consistently reduce enteric CH_4 emissions include nitrates and 3-nitrooxypropanol (3-NOP).

3.1 Nitrate

Nitrate is a form of N which is found naturally within feed, with variable concentrations occurring in different types of forages. Nitrate's potential for reducing enteric CH_4 arises from its ability to act as a [H] sink within the rumen. Both NO_3^- and CO_2 are available as alterative electron acceptors in the rumen and their reduction results in an energy release (ΔG) of 371 and 67 KJ, respectively. Nitrate acts as an alternative electron acceptor, utilising [H] in the reduction of NO_2^- to NH_3 and this reaction is thermodynamically more favourable than the reduction of CO_2 to CH_4 by methanogens (Ungerfeld and Kohn, 2006). Accumulation of NO_2^- within the rumen may occur if the reduction of NO_3^- to NO_2^- takes place at a rate faster than the conversion of NO_2^- to NH_3 (Latham et al., 2016). This accumulation can result in NO_2^- toxicity, reducing the ability of red blood cells to carry oxygen (Lee and Beauchemin, 2014). However, various studies have concluded that provided the rumen microbes are gradually acclimatised to NO_3^-, the risk of toxicity is negligible (Cottle et al., 2011; Olijhoek et al., 2016).

3.1.1 Ruminant studies

Studies have reported a reduction of 16–35% in CH_4 per unit of DM intake across a range of ruminants consuming different diets supplemented with NO_3^-. An increase in total VFA (El-Zaiat et al., 2014; Nolan et al., 2010; van Zijderveld et al., 2010) and NH_3 concentrations (Olijhoek et al., 2016) have also been reported. Though CH_4 production is decreased, no studies have reported an increase in either milk or meat production, despite a predicted increase in the availability of metabolisable energy as a result of a reduction in CH_4 emissions (Johnson and Johnson, 1995). Several studies have reported an accumulation of H_2 within the rumen (Lee et al., 2015c; Olijhoek et al., 2016), a possible outcome of H_2 not being used in the reduction of either CO_2 or NO_3^-. Emission of H_2 also represents a loss of metabolic energy.

Published by Burleigh Dodds Science Publishing Limited, 2020.

The primary factor in the dominance of NO_3^- reduction is related to it being more thermodynamically favourable than the reduction of CO_2 to CH_4; however, NO_3^- also has variable effects on methanogen abundance (Zhou et al., 2014; Liu et al., 2017). Zhao et al. (2018) found that although NO_3^- decreased *in vitro* CH_4 production, rumen fluid collected from donor steers fed NO_3^- exhibited no change in the total methanogen population. However, the relative abundances of *Methanomicrobiales* decreased and *Methanosarcinales* increased with increasing NO_3^- (0%, 1%, 2% DM basis NO_3^-). *Methanosphaera* and *Methanimicrococcus* increased and *Methanoplanus* decreased as a result of inclusion of NO_3^- in the diet. In contrast, Asanuma et al. (2015) found that the populations of methanogens, protozoa and fungi were all drastically decreased when goats were administered 6 or 9 g/d of NO_3^- per day. *Streptococcus bovis and Selenomonas ruminantium* were also increased in this study, possibly suggesting that they may play a role in NO_3^- metabolism (Asanuma et al., 2015).

Though a reduction in enteric CH_4 is consistent with NO_3^- supplementation, this is not often accompanied by a reduction in ruminal methanogen populations. This leads to the conclusion that NO_3^- inhibits that activity of methanogens rather than their growth. There are a large number of bacterial species that may be involved in NO_3^- metabolism. Bacteria identified as potential denitrifiers include *Pseudomonas aeruginosa* and members of *Propionibacterium, Butyrivibrio, Clostridium, Peptostreptococcus, Nitrosomona, Desulfovibrio* and *Enterobacteriaceae* (Latham et al., 2016). In culture, *Selenomonas ruminantium* was tolerant of NO_3^- with some strains exhibiting the ability to reduce NO_3^- to NO_2^-. *Veillonella parvula* and *Wolinella succinogenes* decreased in cultures that lacked NO_3^- and increased with the addition of 5 mM nitrate (Iwamoto et al., 2002).

Nitrite as the intermediate in the reduction of NO_3^- to NH_3 has been shown to cause reductions in cellulolytic bacteria (Iwamoto et al., 2002; Asanuma et al., 2015). However, NO_3^- addition has also been reported to cause no change or an increase in *Fibrobacter succinogenes, Ruminococcus flavefaciens, Ruminococcus albus* as well as total protozoa (Patra and Yu, 2014; Zhao et al., 2015). Protozoal numbers have also been reported to decline with NO_3^- (Nolan et al., 2010; El-Zaiat et al., 2014), which may also contribute to an indirect reduction in CH_4 production as a result of inhibition of commensal methanogens.

3.1.2 Manure

In soils, nitrification is performed by NH_3 oxidising bacteria and archaea, heterotrophic nitrifiers and fungi (Norton, 2008), whereas denitrification is carried out by heterotrophic denitrifiers, denitrifying fungi and autotrophic/heterotrophic nitrifiers (Coyne, 2008). Betaproteobacterial NH_3-oxidising bacteria, *Thaumarchaeota* NH_3-oxidizing archaea and fungi including *Fusarium* and *Trichoderma* (Maeda et al., 2015) are capable of denitrification of NO_3^- to produce N_2O (Maeda et al., 2011).

Compared to a 55% forage and a 45% concentrate diet with encapsulated urea, inclusion of encapsulated slow-release NO_3^- at up to 3% DM linearly decreased total urinary N excretion despite both of the diets being isonitrogenous (Lee et al., 2015a). However, NO_3^--N excretion in urine and faeces was increased and urea-N in the urine was decreased. Decreasing urinary urea-N and increasing urinary NO_3^--N excretion may decrease NH_3 emissions as urea is the primary source of volatile NH_3. However, this could potentially increase direct N_2O emissions as a result of denitrification of NO_3^--N by soil microbes. van Zijderveld et al. (2011) and Li et al. (2012) found that there was no difference in N excretion in dairy cows or sheep when NO_3^- was used to replace urea in the diet. Lee et al. (2015b) established that DM, OM and starch digestibility increased with NO_3^- supplementation, potentially decreasing the amount of substrate that would be available for CH_4 production from manure.

3.2 3-nitrooxypropanol (3-NOP)

Three nitrooxypropanol has been shown to decrease enteric CH_4 emissions by up to 80% (DM intake basis) in beef steers fed a high-grain diet (Vyas et al., 2016). 3-NOP is thought to impede methanogenesis via the inhibition of methyl-coenzyme, which is involved in the transfer of a methyl group to methyl-coenzyme M reductase, the terminal reaction in methanogenesis (Duin et al., 2016).

3.2.1 Ruminant studies

3-NOP has been found to result in a 7.7–80.7% reduction in enteric CH_4 emissions in sheep (Martínez-Fernández et al., 2014), dairy cows (Haisan et al., 2014, 2017; Hristov et al., 2015; Lopes et al., 2016; Reynolds et al., 2014) and beef cattle (Romero-Perez et al., 2014, 2015; Vyas et al., 2016, 2018b). Increasing concentrations of 3-NOP (0.75–4.5 mg NOP/kg BW) have been shown to linearly decrease CH_4 production (from 6.49 to 4.34 CH_4, % gross energy intake) (Romero-Perez et al., 2014).

The energetic saving as a result of reduced CH_4 production has been reported to result in an increase in the body weight of dairy (Haisan et al., 2014; Hristov et al., 2015) and beef steers (Martinez-Fernandez et al., 2018). Others have reported a decrease in the acetate-to-propionate ratio in rumen fluid (Haisan et al., 2014; Romero-Perez et al., 2014, 2015; Martínez-Fernández et al., 2014; Lopes et al., 2016; Haisan et al., 2017), as well as increase in DM digestibility (Haisan et al., 2017) and milk protein (Reynolds et al., 2014).

Published by Burleigh Dodds Science Publishing Limited, 2020.

Though Vyas et al. (2016) found that 3-NOP resulted in a 37.6% decrease in CH_4 and tended to increase the gain-to-feed ratio of beef cattle fed a barley silage-based diet, there was also a tendency for a decrease in average daily gain when these cattle were fed a high-grain finishing diet. Alternatively, Vyas et al. (2018a) found that CH_4 yield decreased and feed conversion efficiency was improved by 3-NOP in cattle-fed both high-forage and high-grain diets.

In pure culture, 3-NOP has also been shown to inhibit the growth of methanogens without affecting other bacteria within the rumen (Duin et al., 2016). The growth and production of CH_4 by *Methanothermobacter marburgensis* was inhibited by both 1 and 10 µM of 3-NOP. However, methanogenesis in the cultures resumed 5 h after administration, suggesting that 3-NOP concentration may have been insufficient to completely inhibit the growth of methanogens. Similarly, growth and methanogenesis by *Methanobrevibacter ruminantium, Methanobacterium smithii, Methanobrevibacter millerae, Methanobacterium bryantii, Methanothermobacter wolfeii, Methanosphaera stadtmanae, Methanomicrobium mobile* and *Methanosarcina barkeri* were all inhibited at 3-NOP concentrations from 0.25 to 10 µM in pure culture (Duin et al., 2016). Not all *in vivo* trials have exhibited a change in the copy number of methanogens as a result of 3-NOP (Romero-Perez et al., 2014; Lopes et al., 2016). Alternatively, Haisan et al. (2014), Romero-Perez et al. (2015) and Martinez-Fernandez et al. (2018) reported a 56.6–64.7% decrease in the abundance of methanogens.

Methanobrevibacter spp and the *Methanomassiliicoccaceae* were reduced by 5.6 and 4.0 fold in Brahman steers fed 2.5 g 3-NOP per day, resulting in a 38% reduction in CH_4 (g/kg DMI). The decrease in *Methanomassiliicoccaceae* may have resulted in the increase of trimethylamines and dimethylamines in the rumen as this family utilises these compounds as substrates for methanogenesis (Martinez-Fernandez et al., 2018). Though originally thought to play a minor role in ruminal methanogenesis, methylotrophic methanogens (*Methanoplasmatales* and *Methanosphaera* spp.) account for 20% of methanogens in the rumen as compared to the 77% that are hydrogenotrophic (mostly *Methanobacteriales*) (Henderson et al., 2015). In beef steers, 3-NOP decreased the ratio of hydrogenotrophic to methylotrophic methanogens, suggesting that this additive has a greater inhibitory effect on hydrogenotrophic methanogens (Martinez-Fernandez et al., 2018).

In sacco, Martinez-Fernandez et al. (2018) found that DM degradability at 24 h was decreased by 7.1%, though DM, OM and neutral detergent fibre (NDF) digestibility was increased in lactating Holstein cows receiving 3-NOP (Haisan et al., 2017). This would imply that the electron carriers produced during fermentation are being used to reduce alternative electron acceptors in the rumen. A 4.9–26.0% increase in propionate production would account for some of this [H] utilization, but increases in propionate have not been observed in all 3-NOP studies (Guyader et al., 2017). In backgrounding beef steers provided with 200 mg/kg BW of 3-NOP, a 37.6% decrease in CH_4 (g/d) coincided with an 89.1% increase in H_2 emissions. Similarly, during the finishing phase, an 84.3% decrease in CH_4 resulted in a 99.8% increase in H_2 (Vyas et al., 2016). Hristov et al. (2015) found a 64-fold increase in H_2 emissions by dairy cows supplemented with 3-NOP, but this accounted for only 3% of the [H] that was spared from methanogenesis (Latham et al., 2016). Alternatively, administration of 3-NOP to steers fed a high-forage diet decreased CH_4 production by 30.2% with no change in H_2 emissions. Propionate concentration was also unchanged; however, the concentration of butyrate was decreased and acetate, isobutyrate, iso-valerate concentrations were increased (Martinez-Fernandez et al., 2018). This suggests that some [H] might have been redirected to microbial cell mass, reflected by the increases in NH_3 and branched-chain VFA.

3.2.2 Manure

Inclusion of 3-NOP resulted in increased faecal DM, NDF and acid detergent fibre (ADF) (Reynolds et al., 2014) and decreased *in sacco* DM digestibility (Martinez-Fernandez et al., 2018). However, it has also been shown to increase DM and NDF digestibility (Haisan et al., 2017). 3-NOP has not been shown to have an impact on CP digestibility or reduce N excretion. This suggests that the relatively small amount of N in 3-NOP is not having an appreciable impact of N metabolism in ruminants. 3-NOP has been reported to rapidly decompose in the environment (Owens et al., unpublished), so it may not be biologically active within manure. Owens et al. (unpublished) investigated the effects of cattle-fed 3-NOP on GHG emissions from composted and stockpiled manure. They concluded that 3-NOP did not affect CO_2, CH_4, N_2O or NH_3 emissions.

Nitrate and 3-NOP have both been shown to suppress enteric CH_4 emissions; however, limited studies have assessed their effects on GHG emissions from a life cycle perspective, including emissions from manure. Concern over NO_2^- poisoning with NO_3 administration has limited its implementation within industry, in spite of studies showing that the reduction of NO_2^- to NH_3 in the rumen can be accelerated through microbial adaptation. Several studies have shown that 3-NOP can dramatically lower enteric CH_4 emissions. However, the magnitude and consistency of this response appears to depend on the forage-to-concentration ratio of the diet. There is a need for appropriate dosage rates to be identified for ruminants-fed different diets so that the appropriate regulatory approval for the use of 3-NOP in ruminants can be obtained. Additionally, determining whether NO_3^- and 3-NOP consistently improve feed efficiency in ruminants is needed to determine if this can serve as an economic incentive for producers to adopt these technologies. Both of these additives also need to be linked with methodologies that enable them to be administered to ruminants in extensive grazing systems, where the intensity of GHG emissions is typically higher than intensive production systems.

4 Plant secondary compounds

A large variety of plant secondary compounds from a diverse range of plants have been explored for their potential to mitigate enteric CH_4 emissions. Secondary metabolic compounds commonly employed as feed additives include essential oils, saponins and tannins. However, over 200 000 defined phytochemicals have been identified (Hartmann, 2007). Tannins are aromatic, recalcitrant compounds with the ability to form complexes with carbohydrates, proteins and minerals that are either reversible or irreversible in nature. They can be classified as hydrolysable or condensed with the most distinguishing feature between these groups being some hydrolysable tannins can be metabolised by rumen microorganisms, whereas condensed tannins (CT) resist biodegradation (Aboagye et al., 2018).

4.1 Tannins

There is considerable evidence that there is an array of tannin-rich forages or tannin extracts which can reduce enteric CH_4 emissions from cattle (Waghorn, 2008; Hess et al., 2006; Grainger et al., 2009; Alves et al., 2017). However, the actual mode of action whereby tannins decrease CH_4 is not well understood. Some hypotheses include a depression of methanogenic archaea, interference with the symbiotic relationship between methanogens and protozoa, inhibition of rumen ciliates (Bhatta et al., 2015), a reduction in fermentable substrates, binding to microbial enzymes (Gonçalves et al., 2011) or acting as a [H] sink (Naumann et al., 2017). CT have a higher binding capacity for dietary proteins and can increase the flow of metabolisable protein to the small intestine, a trait that could decrease microbial protein synthesis. In comparison, hydrolysable tannins exhibit a lower affinity for proteins and bacterial degradation of these tannins may result in the formation of low-molecular-weight metabolites including phenolics which may be toxic to ruminants (Patra and Saxena, 2011).

4.1.1 Ruminant studies

The impact of tannins on ruminal CH_4 emissions has been inconsistent. Grainger et al. (2009) found up to a 30% decrease in CH_4 production when lactating cows were fed up to 266 g of *Acacia mearnsii* per day. Alves et al. (2017) also reported that *Acacia mearnsii* decreased CH_4 emissions (g/kg milk yield) by up to 32% in Holstein dairy cows grazing tropical pastures, with no adverse effects on milk production (yield, fat %, protein %). Koenig et al. (2018) found that addition of

CT from *Acacia mearnsii* to a 40% corn DDGS decreased OM, NDF, ADF, N and gross energy digestibility, with an increase in faecal N (32.4%), and a decrease in urinary N (17.5%). Rumen NH_3-N was also decreased as was VFA concentration and the acetate-to-propionate ratio, but CH_4 production was not measured in this study. Carulla et al. (2005) fed 41 g/kg of a CT extract from *Acacia mearnsii* to growing Swiss White Hill wethers and found that apparent digestibility of OM, CP, NDF and ADF were decreased (1.89-10.6%). Ammonia-N was also decreased but the acetate-to-propionate ratio was increased with propionate increasing by 6.3%. Methane production was decreased as per cent of DM and OM intake, most likely as a result of inhibition of fibre digestibility.

No difference has been observed in performance (feed conversion efficiency) when beef cattle and sheep were fed a diet containing *Acacia mearnsii* CT (Koenig et al., 2018; Carulla et al., 2005). Woodward et al. (2004) found that feeding *Lotus corniculatus*, a forage with CT reduced CH_4 production (dry intake basis) by up to 16%. The CT extract from quebracho trees (*Schinopsis quebracho*) had no effect on CH_4 production and CP digestibility was decreased (Beauchemin et al., 2007). Aboagye et al. (2018) found that a hydrolysable chestnut tannin with and without a CT (quebracho) had no effect on ADG or G:F over 12 weeks in beef cattle. There was only a tendency for a decrease in CH_4 production (g/kg DMI), although digestibility was not reported. These studies highlight the inconsistency of feeding dietary tannins on enteric CH_4 emissions, possibly a result of (i) the tannin-derived species, (ii) phytochemical composition, (iii) administration method and (iv) concentration in the diet.

The ability of tannins to form complexes with rumen microorganisms involves polyphenic reactivity with the cell wall and secreted extracellular enzymes (McSweeney et al., 2001). Bacteria including *Fibrobacter succinogenes*, *Butyrivibrio fibrosolvens*, *Ruminobacter amylophilus* and *Streptococcus bovis* have all been shown to have a high affinity for a diverse range of tannins. McSweeney et al. (1998) and McAllister et al. (1994) suggested that proteolytic bacteria and fungi are less susceptible to tannins than cellulolytic bacteria. Tan et al. (2011) indicated that tannins altered the diversity of methanogens within the rumen when *Leucaena leucocephala* was fed as it decreased the proportion of *Methanomicrobiales* (15.1%) and *Methanobacteriales* (6.8%). In contrast, *Thermoplasmatales* were increased by 21.9% and the overall diversity of the archaeal population was reduced.

The most common response to feeding dietary tannins is a reduction in ruminal CP digestibility. CT form complexes with proteins through H bonds and hydrophobic interactions (Koenig and Beauchemin, 2018). This reduces the availability of dietary protein for microbial degradation within the rumen. It has also been proposed that whilst the protein is bound within the rumen, these complexes dissociate in the abomasum enabling the protein to be digested and the amino acids absorbed in the lower digestive tract. However, this is not

Published by Burleigh Dodds Science Publishing Limited, 2020.

a consistent effect as not all complexes dissociate, often resulting in a reduction in CP digestibility and a shift in N excretion from urine to faeces (Patra and Saxena, 2011).

4.1.2 Manure

Unlike other dietary additives, the effects of including tannins in ruminant diets on manure have been extensively examined (Koenig and Beauchemin, 2018; Powell et al., 2011; Halvorson et al., 2017; Jordan et al., 2015). An advantage of feeding dietary tannins with regard to GHG emissions from manure is that they shift the site of N excretion from urine to faeces. Faecal N in the form of CT-protein complexes is much more stable in manure and less likely to contribute to high NH_3 emissions than urea in urine. Slow-release faecal N is also more likely to be captured by the plant and used for the synthesis of plant proteins.

The inhibitory effect of tannins on urease activity has also been linked to the formation of substrate-tannins complexes (Powell et al., 2011). Tannins may also have the ability to reduce N_2O emissions due to their ability to form complexes with proteins, resulting in insoluble and unavailable forms of N (Powell et al., 2011). Powell et al. (2011) used ventilated chambers to assess the inclusion of dietary tannins on lactating Holstein dairy cow's manure. Cattle were fed a red quebracho (*Schinopsis lorentaii*) and chestnut (*Castanea sativa*) tree mixture at four increasing concentrations (0, 4.5, 9.0 and 18.0 g/kg DMI). Cumulative NH_3 emissions from the tannin slurries were up to 27% lower than the slurry from control animals. Moreover, 54% and 66% of the applied urea was emitted as NH_3 from tannin fed and control slurries, respectively.

As an external application, CT of quebracho were added separately to both composted manure from goats, and N and phosphorus (P) poor soils at 4% (w/w). Tannins decreased cumulative C emissions by 40% and N emissions by 36% in the compost. Tannins applied directly to soil also reduced N_2O emissions by 17%, and reduced NH_3 release by 51% as compared to soil that did not receive tannins (Jordan et al., 2015). Tannins are known to reduce inorganic-N availability by sequestering organic-N sources through complex interactions. Similarly, tannins may act as labile-C sources leading to increased N immobilisation (Kraus et al., 2004). Koenig et al. (2018) fed black wattle at 2.5% with a 40% distillers grain diet to finishing beef steers. An integrated horizontal flux technique with passive NH_3 samplers was used to assess NH_3 emissions from pens. Though measurements were limited, NH_3-N emissions were 23% lower in cattle-fed the diet containing tannins.

Pseudomonas citronellolis and *Pseudomonas plecoglossicidda* were identified as two bacteria capable of utilising tannic (hydrolysable) and gallic acid (phenolic hydrolysate of tannin acid), respectively, within tannery soils.

Bending and Read (1996) found that a hydrolysable polyphenol–protein could be degraded by ectomycorrhizal fungi (*Hysterangium setchellii*, *Lactarius affinis*, *Lactarius controversus*), ericoid mycorrhizal (*Hymenoscyphus ericae*) and wood decomposing fungi (*Hypholoma fasciculare* and *Phanerochaete velutina*) from a soluble tannin acid in forest soils (Mutabaruka et al., 2007). Mutabaruka et al. (2007) found that the ratio of fungi to bacteria increased in systems with high amounts of CT complexes with acidic soils. Though chemically distinct, similar microbes may inhabit tannin manure composts, breaking down tannic complexes during composting or after land application.

Including tannins in ruminant diets has generated varying results on ruminant metabolism and CH_4 emissions. It seems that the CT extracted from black wattle (*Acacia mearnsii*) has been the most consistent at lowering enteric CH_4 emissions. From a manure GHG mitigation perspective, most CT consistently decrease both NH_3 and N_2O emissions. However, tannins can also reduce the digestibility of CP, constituting a loss of N from the animal. Although this could have a negative impact on the productivity of ruminants, it may improve the nutrient composition of manure as a fertiliser and soil amendment.

5 Carbon-derived materials

The cycling of C is defined by the conversion of atmospheric CO_2 to plant biomass-C through photosynthesis. In ruminants, consumed C is partitioned into metabolic animal by-products and waste products, including CO_2 from respiration and the decomposition of manure (Fig. 4). Deposited faeces may decompose and release labile-C, increasing OM levels in surface soil (Sharma et al., 2017). Humic substances (HS) and biochar originally received attention for their ability to increase and sequester soil C. However, more recently, biochar and HS have been assessed for their potential to mitigate enteric CH_4 emissions when they are included directly in the diet of ruminants.

5.1 Humic substances

OM within the soil is a complex heterogeneous mixture of plant- and animal-derived precursors at varying stages of oxidation and decay (Masoom et al., 2016). HS are mainly comprised of polymerized molecules exhibiting strong resistance to biodegradation (Stevenson, 1995). HS can be classified into three operational fractions based on their solubility in alkaline or acidic solutions. Accordingly, fulvic acids are small-sized aliphatic compounds soluble in both alkali and acid, whereas humic acids are high-molecular-weight materials extracted by dilute alkali, that precipitate at a pH of 2 with humins representing the insoluble proportion of HS (Stevenson, 1995; Lamar et al., 2014). The chemical nature of HS may differ widely in terms of functional group structure, composition and reactivity, and consequently influencing their impact on rumen function (Stevenson, 1995).

HS have been postulated to reduce enteric CH_4 production as well as influence zoonotic pathogens. Existing literature has examined the impact of various HS derivatives on both *in vitro* fermentation (Sheng et al., 2017; Terry et al., 2018a; Varadyova et al., 2009) and *in vivo* metabolism (Terry et al., 2018b; El-Zaiat et al., 2018; Ponce et al., 2016). However, there is little consistency between studies, possibly because the concentration and types of HS have not been well characterised.

5.1.1 Ruminant studies

The ability of HS to imitate ionophores was shown when the average daily gain, DM intake and feed efficiency of cattle administered monensin did not differ from those supplemented with HS (McMurphy et al., 2009). However, as no negative control was utilised in this study, results need to be interpreted with caution. Saanen goats fed diets containing humic acids at up to 3% diet DM had higher milk yields and lower levels of blood cholesterol (Degirmenci, 2012). Agazzi et al. (2007) found that milk consumption and average daily gain in newborn kids fed HS were higher than the control group. This study hypothesised that it was the antibiotic properties of HS that improved cell-mediated immunity, decreasing the risk of digestive orders and diarrhoea in young kids.

Recently, HS have been evaluated for their ability to inhibit ruminal methanogenesis. Whilst an *in vitro* batch culture showed that HS resulted in a consistent decrease in CH_4 production (Sheng et al., 2017), this was not verified in continuous culture using the rumen simulation technique (Rusitec) (Terry et al., 2018a). Further to this, a study conducted to investigate the effect of HS on beef heifers fed a barely silage-based diet found that there was no effect on enteric CH_4 production (Terry et al., 2018b). However, there was an increase in total N retention when heifers were fed up to 300 mg/kg live BW of HS, suggesting an improvement in protein utilisation.

Both Terry et al. (2018a,b) examined the rumen microbial population using 16s rRNA sequencing. In the Rusitec, *Fibrobacter* and *Christensenellaceae* R-7 were reduced by HS in solid-associated samples. Although the *Methanobacteria* were not changed by HS, *Methanobacterium* was increased and *Methanobrevibacter* and *Methanosphaera* were decreased. *In vivo*, the relative abundance of *Proteobacteria*, *Synergistetes* and *Euryarchaeota* were decreased by HS.

5.1.2 Manure

Information concerning the incorporation of HS into manure mixtures is scarce. In this regard, HS were shown to decrease total N excretion when fed to beef heifers. This may imply that HS could have the potential to reduce NH_3 emissions from manure. This is in agreement with the findings of Shi et al. (2001) who investigated the effects of black and brown humates on NH_3 emissions after their inclusion in a soil-faeces-urine mixture (1.7% of total mixture mass). Results indicated that NH_3 emissions decreased by 39.8% compared to the manure-soil mixture with no humates. However, in this experiment HS were added to the manure after excretion, so results may differ from when HS are included directly in the diet.

Various studies have reported that HS decrease (Miller et al., 2015; Blodau and Deppe, 2012; Tan et al., 2018) CH_4 emissions under soil anoxic environments. This is thought to be induced by functional structures within HS that can act as electron acceptors (i.e. hydrogen during methanogenesis) (Martinez et al., 2013; Terry et al., 2018b). However, there is also contradictory evidence which found that HS facilitated CH_4 production in anoxic paddy soils (Zhou et al., 2014).

5.2 Biochar

Biochar, a pyrolysed thermal degraded form of black-C, is obtained by heating (350–600°C) plant biomass residues under oxygen-limited conditions (Cha et al., 2016). Biochar is mostly comprised of recalcitrant-C, but also contains an array of inorganic nutrients (Joseph et al., 2018). In general, biochar is characterised by a porous structure, large surface area and high mineral content (Cha et al., 2016), characteristics which depend on the original biomass or feedstock source. Similarly, other reports have found biochar to exhibit high ion exchange and absorption properties (Yuan et al., 2017) that make it suitable as a soil amendment, water and air scrubber and a detoxifying agent (Tawheed and Baowei, 2017). The ability of biochar to reduce N_2O and CH_4 emissions in cultivated fields (Karhu et al., 2011; Cayuela et al., 2014) and act as a detoxifying agent has sparked interest in their use as a feed additive for ruminants.

5.2.1 Ruminant studies

The original proposal for using biochar as a dietary CH_4 mitigation tool comes from Leng et al. (2012), who hypothesised that, due to its porous nature, biochar may promote the formation of biofilms or induce interspecies electron transfer within the rumen. Additionally, biochar has been hypothesised to increase the population of CH_4-oxidising bacteria, methanotrophs, within the rumen; even though it has been shown in vitro that methanotrophs account for a small proportion of microbial activity (Kajikawa et al., 2003). Absorption of CH_4 by biochar was also thought to play a significant role in the reduction of CH_4 production. However, as described by Saleem et al. (2018), it seems unlikely that biochar (typically added at 0.5–2% of diet DM) would absorb the large volume of CH_4 that is typically produced by rumen methanogens.

Leng et al. (2012) used 12 'yellow' calves from Laos to investigate the effect of both nitrate and biochar produced from rice husks on CH_4 production and growth performance. Methane production was reduced by 22% with biochar and live weight gain was increased by an almost unbelievable 25%. However, these results need to be interpreted with caution as they did not use continuous calorimetry for CH_4 measurements. Saleem et al. (2018) examined a pine-based biochar using the Rusitec and found that CH_4 (g/g DM digested) was decreased by 22.4% when biochar was fed at up to 2.0% of diet DM. Similarly, total VFA, NH_3-N and nutrient disappearance of DM, OM, CP, NDF and ADF were all improved. In contrast, using the same product, Terry et al. (2019) found that VFA, CH_4 production and apparent digestibility were not affected by biochar when it comprised up to 2.0% of dietary DM of a barley silage-based diet fed to beef heifers.

5.2.2 Manure

Joseph et al. (2015) found that the manure of cows fed a mixture of molasses (0.1 kg day^{-1}) and high temperature jarrah wood (*Eucalyptus marginata*) biochar (0.33 kg day^{-1}) improved soil properties and increased OM sequestration (0–40 cm) in an Australian Chromosol. This response was attributed to enhanced biochar N and P adsorption from the cow's gut and limited transformation of recalcitrant-C upon digestion, increasing stable C which enhanced soil fertility.

Yuan et al. (2017) employed rice (Oryza sativa) husk-derived biochar as co-compositing element for chicken manure. Compared to compost, biochar reduced soil CO_2 and N_2O emissions by 35% and 27%, respectively. They hypothesised that biochar increased OM stabilisation through the soil profile and impacted denitrifying bacterial populations as evidenced by an increase in the archaeal genes encoding for enzymes related to bacterial nitrification. The abundance of 16S rRNA was decreased in biochar amended manure. Similarly, biochar has been shown to effectively retain NH_3 and N_2O in co-composted poultry litter (Steiner et al., 2010). Jia et al. (2016) found rice hull biochar decreased the peak rate of N_2O emissions by 60% compared to pure compost when it was used as a bulking agent.

From available research (Atkinson et al., 2010; Jia et al., 2016; Steiner et al., 2010; Yuan et al., 2017), it seems that biochar amended compost is an effective mitigation strategy for composted chicken manure. More research is required to evaluate whether the same responses occur with composted beef cattle manure. It will also be important to evaluate how feeding biochar to ruminants alters post-excretion GHG emissions from manure.

From current research, it can be concluded that HS and biochar are largely ineffective at mitigating ruminal CH_4 production. Although HS did not mitigate CH_4 production in ruminants, there may still be associated health or metabolic benefits as shown by an increase in average daily gain in goats (Agazzi et al., 2007) and N retention in beef heifers (Terry et al., 2018b). However, assessment using a performance trial is needed to support this evaluation. Although biochar has been shown to mitigate enteric CH_4 *in vitro*, this response has not been confirmed *in vivo*. The current mechanisms proposed as to how biochar may mitigate enteric CH_4 are not supported by current *in vivo* findings, and further assessment of how it may alter the rumen microbiome is required. Limited studies have suggested that both HS and biochar are successful at mitigating manure emissions, and perhaps at defined levels, feeding these additives will improve the nutrient composition of manure as a fertilizer without negative impacts on ruminant performance.

6 Microbial hydrogen utilisation

It is generally accepted that an increase in the partial pressure of [H] within the rumen will result in an inhibition of fermentation through reduced re-oxidisation of co-factors (Ungerfeld, 2015b). Additionally, stoichiometry shows that a decrease in enteric CH_4 production from ruminants should result in more energy for maintenance and production (Johnson and Johnson, 1995). However, these two concepts are not always observed as responses to a reduction in enteric CH_4 production.

Hydrogenotrophic archaea (*Methanobrevibacter, Methanobacterium*) are the predominant archaea (Henderson et al., 2015) and can utilise [H] or to a much lesser extent formate, as sources of electrons to reduce CO_2 to CH_4 (Richards et al., 2016). The contribution of formate to CH_4 production is estimated at 18% of CH_4 produced in the rumen (Tapio et al., 2017b). Methylotrophic methanogens (i.e. *Methanosarcinales, Methanosphaera* and *Methanomassiliicoccaceae*) are less abundant and can utilise methanol and methylamines to produce CH_4 (Huws et al., 2018). The acetoclastic pathway can also result in the formation of CH_4 from acetate, but the *Methanosarcinales* spp. which utilise this pathway have a slow growth rate and are not prominent within the mature rumen (Friedman et al., 2017b). Whilst the majority of [H] is utilised by archaea, several other means of [H] disposal may occur in the rumen, including the use of reducing equivalents to reduce sulphate and nitrate (NO_3^-) as well as reductive acetogenesis, propionogenesis and the synthesis of microbial biomass.

Sulphate and NO_3^- reduction reactions are more thermodynamically favourable than the reduction of CO_2 (Morgavi et al., 2010; Haque, 2018) and sulphate and NO_3^- reducing bacteria have been shown to outcompete methanogens in anoxic environments (Scheid et al., 2003). Reductive acetogenesis is not as thermodynamically favourable as methanogenesis and it is unlikely that the rumen would establish a dominant and sustained microbial population capable of this process (Fonty et al., 2007; Friedman et al., 2017b). However, if a reductive acetogen was developed that could exist naturally in the rumen and utilise only [H], rather than sugars (obligate hydrogenotroph), it could potentially act as a successful [H] sink and suppress methanogenesis (Ungerfeld, 2015a).

Feed additives which redirect [H] towards an alternative metabolic sink represent a new avenue for investigation. Martinez-Fernandez et al. (2017) supplemented Brahman steers with the antimethanogenic compound, chloroform. Half of the steers were also administered phloroglucinol, an intermediate metabolite of flavonoid degradation which through the utilisation [H] and formate can form acetate. Phloroglucinol resulted in an increase in acetate production, *Prevotella*, *Ruminococcus* and *Fibrobacter* abundance as well as a decrease in H_2 and formate production. This study was the first *in vivo* trial to demonstrate that [H] can be redirected towards the reduction phloroglucinol as a means of inhibiting methanogenesis. Further studies examining the redirection of [H] during the inhibition of methanogenesis, and how this alters the rumen microbiome are needed. For example, 3-NOP can successfully decrease methanogenesis; however, excess [H] is not captured through other reductive process and the H_2 emission is increased. Redirection of this [H] into a usable metabolite would result in a further reduction of CH_4 production and CH_4 intensity, potentially without the loss of energy in the form of H_2.

It has been suggested that, as within other anaerobic environments, there is a balance between methane producing and methane utilising microbes. Methanotrophs are specific archaea or bacteria which can metabolise CH_4 in the presence of oxygen (Leng, 2014). However, CH_4 can also be anaerobically oxidised utilising existing oxygen within sulphate, metal oxides and nitrate (Joye, 2012). Their presence or importance in the ruminal environment is a matter of debate.

In an artificial rumen system, it was found that only 0.2–0.5% of CH_4 produced was oxidised by coupling with sulphate reduction (Kajikawa et al., 2003). However, it is theorised that methanotrophs are more likely to colonise the rumen wall due to diffusion of oxygen from the bloodstream. A metagenomic analysis of the microbial populations in beef cattle rumen did not detect methanotrophs (Wallace et al., 2015). Using methanotroph-specific primers, Mitsumori et al. (2002) suggested that methanotrophs exist in both ruminal fluid and the biofilm attached to the rumen wall, but only type I methanotrophs were detected. Type I methanotrophs include *Methylomonas*, *Methylobacter*, *Methylomicrobium* and *Methylococcus* which utilise the ribulose monophosphate pathway to assimilate carbon. Type II includes *Methylocystis* and *Methylosinus*, which use the serine pathway to assimilate carbon (Mitsumori et al., 2002). Jin et al.

(2017) found that the *Methylococcaceae* family was dominant in solid, liquid and rumen wall-associated populations. In rumen batch culture, Liu et al. (2017) found that the addition of NO_3^- decreased methanogenesis and increased the phylum NC10. The NC10 bacteria are the only known bacteria that are capable of anaerobic oxidisation of CH_4. The bacterium *Methoxymirabilis oxyfera* converts NO_2^- to nitric oxide and then dismutates nitric oxide into nitrogen and oxygen, using the resulting O_2 to support CH_4 oxidation (He et al., 2016; Joye, 2012). All other microbes with the ability to anaerobically oxidise CH_4 are archaea. Klieve et al. (2012) identified that rumen contents of cattle had mcrA gene sequences relating to CH_4 oxidising archaea, represented by archaea which have been shown to anaerobically catabolise methane using sulphate reduction in sediments from the Gulf of Mexico (Lloyd et al., 2006).

Parmar et al. (2015) established that Type I methanotrophs were more abundant in 50:50 forage-to-concentrate diets, whereas Type II increased in a complete forage diet. The enzyme formate dehydrogenase, which oxidises formate, was increased within the high-forage diet, a finding consistent with an increase in Type II methanotrophs. Auffret et al. (2018) also identified the presence of three methanotrophic bacteria including *Methylobacterium*, *Methylomonas* and *Methylomicrobium* in low abundance ($0.1 \pm 0.01\%$) in the rumen of beef steers. The *Methylomonas* were more abundant and negatively correlated with CH_4 emissions. The overall diversity of methanotrophs was greater in high CH_4 emitters as compared to low emitters. The inconsistency in the identification of methanotrophs in the rumen could reflect the lack of sequencing depth and breadth for these rare populations.

The importance of methanotrophs in a nutrient-rich environment like the rumen is questionable. However, the use of CH_4 by methanotrophs may partially account for [H] that is not stoichiometrically accounted for when CH_4 inhibitors such as NO_3^- and 3-NOP are included in ruminant diets. Methanotrophs are important in other anaerobic environments such as sediments, the oceanic seafloor and both freshwater and saline water systems (He et al., 2016), oxidising over 80% of the emitted CH_4 before it reaches the atmosphere (Cai et al., 2016). However, these are relatively stable environments, and do not experience the same passage rate or daily variation in substrate availability as within the rumen. As sequencing technologies improve our ability to delve deeper into the ruminal microbiome, a more detailed identification of methanotrophs should enhance our understanding of [H] balance in the rumen.

7 Future trends and conclusion

Presently, 3-NOP and NO_3^- can mitigate enteric CH_4 production while having little effects on GHG emissions from manure; however, the excess [H] is not completely captured in the form of reduced substrates (Table 1). Likewise,

Table 1 Summary of dietary additives and their implications for GHG mitigation from ruminant production

Dietary additive	Enteric emissions	Improvement in product	Manure emissions	Improvement in product	Interaction	Recommend
Nitro compounds						
Nitrate	$\downarrow CH_4$ $\uparrow H_2$	No	N/A[a], may $\uparrow N_2O$, $\downarrow NH_3$	N/A	N/A, Likely	Yes[a]
3 – NOP	$\downarrow CH_4$ $\uparrow H_2$	Inconsistent	N/A	N/A	N/A, Unlikely	Yes[a]
Secondary compounds						
Tannins	Variable, may $\downarrow CH_4$	Inconsistent, may $\downarrow DMI$	$\downarrow NO_2$ $\downarrow NH_3$	Yes	Yes	Too variable
Organic carbon						
Humic substances	No	N/A	N/A[a], may $\downarrow NH_3$	N/A[a], may $\uparrow$ stable C	No	N/A
Biochar	No	N/A	N/A[a] may $\downarrow NO_2$, $\downarrow NH_3$	N/A[a], may $\uparrow$ stable C	N/A	N/A

[a] Based on limited research.
N/A = information not available.

tannins can reduce GHG emissions from manure, whereas their effect on enteric CH_4 emissions from ruminants is highly variable. Organic C additives may have potential for mitigation of manure GHG, but there is limited research to support their ability to reduce enteric CH_4 emissions.

Dietary manipulation as a mitigation strategy is thought to be the most viable method for reducing GHG emissions from ruminants. However, as highlighted, there is a balance to be met towards ensuring disrupting rumen metabolism does not cause unintended increases in GHG from manure. There are also additional considerations for how dietary changes alter the rumen microbiome and how long these changes are sustained. Investigations regarding the effects of dietary additives on both enteric and manure CH_4 emissions have reinforced the complexity of the dynamics between enteric- and manure-CH_4. Though enteric CH_4 production has a much larger CO_2-equivalent contribution to total GHG emissions, N_2O produced by manure is a more potent GHG. Therefore, when recommending GHG mitigation strategies from ruminants, it is important to validate its efficiency at a whole-farm level.

8 Where to look for further information

8.1 Further reading

Extensive review of GHG mitigation strategies from livestock production: 'Mitigation of Greenhouse gas emissions in livestock production – FAO Animal production and Health'. Available at: http://www.fao.org/docrep/018/i3288e/i3288e00.htm.

Extensive characterisation of livestock production by region and associated GHG production: 'Assessment of greenhouse gas emissions and mitigation potential' – FAO Global Livestock Environmental Assessment Model (GLEAM). Available at: http://www.fao.org/gleam/results/en/.

A review on agricultural N cycle: Robertson and Vitousek (2009).

8.2 Key journals/conferences

The Greenhouse Gas and Animal Agriculture (GGAA) Conference is an international conference held every 3 years.

8.3 Major international research projects

Biochar Project: Assessment of the potential for adding biochar to beef cattle diets to reduce GHG emissions in agriculture: http://www.agr.gc.ca/eng/programs-and-services/agricultural-greenhouse-gases-program/approved-projects/?id=1508423883267.

9 References

Aboagye, I. A., Oba, M., Castillo, A. R., Koenig, K. M., Iwaasa, A. D. and Beauchemin, K. A. 2018. Effects of hydrolyzable tannin with or without condensed tannin on methane emissions, nitrogen use, and performance of beef cattle fed a high-forage diet. *J. Anim. Sci.* 96(12), 5276–86. doi:10.1093/jas/sky352.

Adler, A. A., Doole, G. J., Romera, A. J. and Beukes, P. C. 2015. Managing greenhouse gas emissions in two major dairy regions of New Zealand: a system-level evaluation. *Agric. Sys.* 135, 1–9. doi:10.1016/j.agsy.2014.11.007.

Agazzi, A., Cigalino, G., Mancin, G., Savoini, G. and Dell'Orto, V. 2007. Effects of dietary humates on growth and an aspect of cell-mediated immune response in newborn kids. *Small Rumin. Res.* 72(2–3), 242–5. doi:10.1016/j.smallrumres.2006.10.020.

Alves, T. P., Dall-Orsoletta, A. C. and Ribeiro-Filho, H. M. N. 2017. The effects of supplementing *Acacia mearnsii* tannin extract on dairy cow dry matter intake, milk production, and methane emission in a tropical pasture. *Trop. Anim. Health Prod.* 49(8), 1663–8. doi:10.1007/s11250-017-1374-9.

Asanuma, N., Yokoyama, S. and Hino, T. 2015. Effects of nitrate addition to a diet on fermentation and microbial populations in the rumen of goats, with special reference to *Selenomonas ruminantium* having the ability to reduce nitrate and nitrite. *Anim. Sci. J.* 86(4), 378–84. doi:10.1111/asj.12307.

Atkinson, C. J., Fitzgerald, J. D. and Hipps, N. A. 2010. Potential mechanisms for achieving agricultural benefits from biochar application to temperate soils: a review. *Plant Soil* 337(1–2), 1–18. doi:10.1007/s11104-010-0464-5.

Auffret, M. D., Stewart, R., Dewhurst, R. J., Duthie, C. A., Rooke, J. A., Wallace, R. J., Freeman, T. C., Snelling, T. J., Watson, M. and Roehe, R. 2018. Identification, comparison, and validation of robust rumen microbial biomarkers for methane emissions using diverse *Bos taurus* breeds and basal diets. *Front. Microbiol.* 8, 2642. doi:10.3389/fmicb.2017.02642.

Bach, A., Calsamiglia, S. and Stern, M. D. 2005. Nitrogen metabolism in the rumen. *J. Dairy Sci.* 88 (Suppl. 1), E9–21. doi:10.3168/jds.S0022-0302(05)73133-7.

Beauchemin, K. A., McGinn, S. M., Martinez, T. F. and McAllister, T. A. 2007. Use of condensed tannin extract from quebracho trees to reduce methane emissions from cattle. *J. Anim. Sci.* 85(8), 1990–6. doi:10.2527/jas.2006-686.

Beauchemin, K. A., Kreuzer, M., O'Mara, F. and McAllister, T. A. 2008. Nutritional management for enteric methane abatement: a review. *Aust. J. Exp. Agric.* 48(2), 21–7. doi:10.1071/EA07199.

Belanche, A., Doreau, M., Edwards, J. E., Moorby, J. M., Pinloche, E. and Newbold, C. J. 2012. Shifts in the rumen microbiota due to the type of carbohydrate and level of protein ingested by dairy cattle are associated with changes in rumen fermentation. *J. Nutr.* 142(9), 1684–92. doi:10.3945/jn.112.159574.

Bending, G. D. and Read, D. J. 1996. Nitrogen mobilization from protein-polyphenol complex by ericoid and ectomycorrhizal fungi. *Soil Biol. Biochem.* 28(12), 1603–12. doi:10.1016/S0038-0717(96)00258-1.

Bhatta, R., Saravanan, M., Baruah, L. and Prasad, C. S. 2015. Effects of graded levels of tannin-containing tropical tree leaves on *in vitro* rumen fermentation, total protozoa and methane production. *J. Appl. Microbiol.* 118(3), 557–64. doi:10.1111/jam.12723.

Blodau, C. and Deppe, M. 2012. Humic acid addition lowers methane release in peats of the Mer Bleue bog, Canada. *Soil Biol. Biochem.* 52, 96–8. doi:10.1016/j.soilbio.2012.04.023.

Boots, B., Lillis, L., Clipson, N., Petrie, K., Kenny, D. A., Boland, T. M. and Doyle, E. 2013. Responses of anaerobic rumen fungal diversity (phylum Neocallimastigomycota) to changes in bovine diet. *J. Appl. Microbiol.* 114(3), 626–35. doi:10.1111/jam.12067.

Broderick, G. A. 2003. Effects of varying dietary protein and energy levels on the production of lactating dairy cows. *J. Dairy Sci.* 86(4), 1370–81. doi:10.3168/jds.S0022-0302(03)73721-7.

Cai, Y., Zheng, Y., Bodelier, P. L. E., Conrad, R. and Jia, Z. 2016. Conventional methanotrophs are responsible for atmospheric methane oxidation in paddy soils. *Nat. Commun.* 7, 11728. doi:10.1038/ncomms11728.

Carulla, J. E., Kreuzer, M., Machmüller, A. and Hess, H. D. 2005. Supplementation of *Acacia mearnsii* tannins decreases methanogenesis and urinary nitrogen in forage-fed sheep. *Aust. J. Agric. Res.* 56(9), 961–70. doi:10.1071/AR05022.

Castillo-Lopez, E., Ramirez Ramirez, H. A., Klopfenstein, T. J., Anderson, C. L., Aluthge, N. D., Fernando, S. C., Jenkins, T. and Kononoff, P. J. 2014. Effect of feeding dried distillers grains with solubles on ruminal biohydrogenation, intestinal fatty acid profile, and gut microbial diversity evaluated through DNA pyro-sequencing. *J. Anim. Sci.* 92(2), 733–43. doi:10.2527/jas.2013-7223.

Castillo-Lopez, E., Jenkins, C. J. R., Aluthge, N. D., Tom, W., Kononoff, P. J. and Fernando, S. C. 2017. The effect of regular or reduced-fat distillers grains with solubles on rumen methanogenesis and the rumen bacterial community. *J. Appl. Microbiol.* 123(6), 1381–95. doi:10.1111/jam.13583.

Cayuela, M. L., van Zwieten, L., Singh, B. P., Jeffery, S., Roig, A. and Sánchez-Monedero, M. A. 2014. Biochar's role in mitigating soil nitrous oxide emissions: a review and meta-analysis. *Agric. Ecosyst. Environ.* 191, 5–16. doi:10.1016/j.agee.2013.10.009.

Cha, J. S., Park, S. H., Jung, S.-C., Ryu, C., Jeon, J.-K., Shin, M. and Park, Y. 2016. Production and utilization of biochar: a review. *J. Ind. Eng. Chem.* 40, 1–15. doi:10.1016/j. jiec.2016.06.002.

Cottle, D. J., Nolan, J. V. and Wiedemann, S. G. 2011. Ruminant enteric methane mitigation: a review. *Anim. Prod. Sci.* 51(6), 491–514. doi:10.1071/AN10163.

Coyne, M. S. 2008. Biological denitrification. In: Schepers, J. S. and Raun, W. R. (Eds), *Nitrogen in Agricultural Systems.* American Society of Agronomy, Crop Science Society of America, Soil Science Society of America, Madison, WI, pp. 201–53.

Degirmenci, T. 2012. Effects of diets containing humic acid on the milk yield, milk composition and blood metabolites in Saanen goats. *Res. J. Anim. Sci.* 6(1), 4–7. doi:10.3923/rjnasci.2012.4.7.

Dijkstra, J., Oenema, O., van Groenigen, J. W., Spek, J. W., van Vuuren, A. M. and Bannink, A. 2013. Diet effects on urine composition of cattle and N_2O emissions. *Animal* 7, 292–302. doi:10.1017/S1751731113000578.

Duin, E. C., Wagner, T., Shima, S., Prakash, D., Cronin, B., Yáñez-Ruiz, D. R., Duval, S., Rümbeli, R., Stemmler, R. T., Thauer, R. K. and Kindermann, M. 2016. Mode of action uncovered for the specific reduction of methane emissions from ruminants by the small molecule 3-nitrooxypropanol. *Proc. Natl. Acad. Sci. U.S.A.* 113(22), 6172–7. doi:10.1073/pnas.1600298113.

El-Zaiat, H. M., Araujo, R. C., Soltan, Y. A., Morsy, A. S., Louvandini, H., Pires, A. V., Patino, H. O., Correa, P. S. and Abdalla, A. L. 2014. Encapsulated nitrate and cashew nut shell liquid on blood and rumen constituents, methane emission, and growth performance of lambs. *J. Anim. Sci.* 92(5), 2214–24. doi:10.2527/jas.2013-7084.

El-Zaiat, H. M., Morsy, A. S., El-Wakeel, E. A., Anwer, M. M. and Sallam, S. M. 2018. Impact of humic acid as an organic additive on ruminal fermentation constituents, blood parameters and milk production in goats and their kids growth rate. *J. Anim. Feed Sci.* 27(2), 105–13. doi:10.22358/jafs/92074/2018.

Enjalbert, F., Combes, S., Zened, A. and Meynadier, A. 2017. Rumen microbiota and dietary fat: a mutual shaping. *J. Appl. Microbiol.* 123(4), 782–97. doi:10.1111/jam.13501.

Erisman, J. W., Galloway, J. N., Seitzinger, S., Bleeker, A., Dise, N. B., Petrescu, A. M., Leach, A. M. and de Vries, W. 2013. Consequences of human modification of the global nitrogen cycle. *Philos. Trans. R. Soc. Lond., B, Biol. Sci.* 368(1621), 20130116. doi:10.1098/rstb.2013.0116.

Fageria, N. K. and Baligar, V. C. 2005. Enhancing nitrogen use efficiency in crop plants. In: *Advances in Agronomy.* Academic Press, pp. 97–185. doi:10.1016/S0065-2113(05)88004-6.

Firkins, J. L. and Yu, Z. 2015. RUMINANT NUTRITION SYMPOSIUM: how to use data on the rumen microbiome to improve our understanding of ruminant nutrition. *J. Anim. Sci.* 93(4), 1450–70. doi:10.2527/jas.2014-8754.

Fonty, G., Joblin, K., Chavarot, M., Roux, R., Naylor, G. and Michallon, F. 2007. Establishment and development of ruminal hydrogenotrophs in methanogen-free lambs. *Appl. Environ. Microbiol.* 73(20), 6391–403. doi:10.1128/AEM.00181-07.

Friedman, N., Shriker, E., Gold, B., Durman, T., Zarecki, R., Ruppin, E. and Mizrahi, I. 2017a. Diet-induced changes of redox potential underlie compositional shifts in the rumen archaeal community. *Environ. Microbiol.* 19(1), 174–84. doi:10.1111/1462-2920.13551.

Friedman, N., Jami, E. and Mizrahi, I. 2017b. Compositional and functional dynamics of the bovine rumen methanogenic community across difrerent developmental stages. *Environ. Microbiol.* 19(8), 3365–73. doi:10.1111/1462-2920.13846.

Galloway, J. N., Dentener, F. J., Capone, D. G., Boyer, E. W., Howarth, R. W., Seitzinger, S. P., Asner, G. P., Cleveland, C. C., Green, P. A., Holland, E. A., Karl, D. M., Michaels, A. F., Porter, J. H., Townsend, A. R. and Vöosmarty, C. J. 2004. Nitrogen cycles: past, present, and future. *Biogeochemistry* 70(2), 153–226. doi:10.1007/s10533-004-0370-0.

Gautam, D. P., Rahman, S., Borhan, M. S. and Engel, C. 2016. The effect of feeding high fat diet to beef cattle on manure composition and gaseous emission from a feedlot pen surface. *J. Anim. Sci. Technol.* 58, 22–. doi:10.1186/s40781-016-0104-6.

Gerber, P. J., Steinfeld, H., Henderson, B., Mottet, A., Opio, C., Dijkman, J., Falcucci, A. and Tempio, G. 2013. *Tackling Climate Change through Livestock: a Global Assessment of Emissions and Mitigation Opportunities.* Food and Agriculture Organization of the United Nations, Rome.

Gonçalves, R., Mateus, N. and de Freitas, V. 2011. Inhibition of ⍺-amylase activity by condensed tannins. *Food Chem.* 125(2), 665–72. doi:10.1016/j.foodchem.2010.09.061.

Grainger, C., Clarke, T., Auldist, M. J., Beauchemin, K. A., McGinn, S. M., Waghorn, G. C. and Eckard, R. J. 2009. Potential use of *Acacia mearnsii* condensed tannins to reduce methane emissions and nitrogen excretion from grazing dairy cows. *Can. J. Anim. Sci.* 89(2), 241–51. doi:10.4141/CJAS08110.

Griffin, W. A., Bremer, V. R., Klopfenstein, T. J.,ʻStalker, L. A., Lomas, L. W., Moyer, J. L. and Erickson, G. E. 2012. A meta-analysis evaluation of supplementing dried distillers grains plus solubles to cattle consuming forage-based diets 1. *Prof. Anim. Sci.* 28(3), 306–12. doi:10.15232/S1080-7446(15)30360-0.

Guyader, J., Eugene, M., Noziere, P., Morgavi, D. P., Doreau, M. and Martin, C. 2014. Influence of rumen protozoa on methane emission in ruminants: a meta-analysis approach. *Animal* 8(11), 1816–25. doi:10.1017/S1751731114001852.

Guyader, J., Ungerfeld, E. M. and Beauchemin, K. A. 2017. Redirection of metabolic hydrogen by inhibiting methanogenesis in the rumen simulation technique (RUSITEC). *Front. Microbiol.* 8, 393. doi:10.3389/fmicb.2017.00393.

Haisan, J., Sun, Y., Guan, L. L., Beauchemin, K. A., Iwaasa, A., Duval, S., Barreda, D. R. and Oba, M. 2014. The effects of feeding 3-nitrooxypropanol on methane emissions and productivity of Holstein cows in mid lactation. *J. Dairy Sci.* 97(5), 3110–9. doi:10.3168/jds.2013-7834.

Haisan, J., Sun, Y., Guan, L., Beauchemin, K. A., Iwaasa, A., Duval, S., Kindermann, M., Barreda, D. R. and Oba, M. 2017. The effects of feeding 3-nitrooxypropanol at two doses on milk production, rumen fermentation, plasma metabolites, nutrient digestibility, and methane emissions in lactating Holstein cows. *Anim. Prod. Sci.* 57(2), 282–9. doi:10.1071/AN15219.

Hales, K. E., Cole, N. A. and Macdonald, J. C. 2013. Effects of increasing concentrations of wet distillers grains with solubles in steam-flaked, corn-based diets on energy metabolism, carbon-nitrogen balance, and methane emissions of cattle. *J. Anim. Sci.* 91(2), 819–28. doi:10.2527/jas.2012-5418.

Hall, M. B. and Huntington, G. B. 2008. Nutrient synchrony: sound in theory, elusive in practice. *J. Anim. Sci.* 86(14 Suppl), E287–92. doi:10.2527/jas.2007-0516.

Halvorson, J. J., Kronberg, S. L. and Hagerman, A. E. 2017. Effects of dietary tannins on total and extractable nutrients from manure. *J. Anim. Sci.* 95(8), 3654–65. doi:10.2527/jas.2016.1129.

Haque, M. N. 2018. Dietary manipulation: a sustainable way to mitigate methane emissions from ruminants. *J. Anim. Sci. Technol.* 60, 15–. doi:10.1186/s40781-018-0175-7.

Hartmann, T. 2007. From waste products to ecochemicals: fifty years research of plant secondary metabolism. *Phytochemistry* 68(22–24), 2831–46. doi:10.1016/j.phytochem.2007.09.017.

He, Z., Cai, C., Wang, J., Xu, X., Zheng, P., Jetten, M. S. and Hu, B. 2016. A novel denitrifying methanotroph of the NC10 phylum and its microcolony. *Sci. Rep.* 6, 32241. doi:10.1038/srep32241.

Henderson, G., Cox, F., Ganesh, S., Jonker, A., Young, W., Global Rumen Census Collaborators and Janssen, P. H. 2015. Rumen microbial community composition varies with diet and host, but a core microbiome is found across a wide geographical range. *Sci. Rep.* 5, 14567. doi:10.1038/srep14567.

Hess, H. D., Tiemann, T. T., Noto, F., Carulla, J. E. and Kreuzer, M. 2006. Strategic use of tannins as means to limit methane emission from ruminant livestock. *Int. Congr. Ser.* 1293, 164–7. doi:10.1016/j.ics.2006.01.010.

Hook, S. E., Steele, M. A., Northwood, K. S., Wright, A. D. and McBride, B. W. 2011. Impact of high-concentrate feeding and low ruminal pH on methanogens and protozoa in the rumen of dairy cows. *Microb. Ecol.* 62(1), 94–105. doi:10.1007/s00248-011-9881-0.

Hristov, A. N., Callaway, T. R., Lee, C. and Dowd, S. E. 2012. Rumen bacterial, archaeal, and fungal diversity of dairy cows in response to ingestion of lauric or myristic acid. *J. Anim. Sci.* 90(12), 4449–57. doi:10.2527/jas.2011-4624.

Hristov, A. N., Oh, J., Lee, C., Meinen, R., Montes, F., Ott, T., Firkins, J., Rotz, A., Dell, C., Adesogan, A., Yang, W., Tricarico, J., Kebreab, E., Waghorn, G., Dijkstra, J. and Oosting, S. 2013. *Mitigation of Greenhouse Gas Emissions in Livestock Production – A Review of Technical Options for non-CO2 Emissions*. Gerber, P. J., Henderson, B. and Makkar, H. P. S. (Eds). Food and Agriculture Organization, Rome, Italy.

Hristov, A. N., Oh, J., Giallongo, F., Frederick, T. W., Harper, M. T., Weeks, H. L., Branco, A. F., Moate, P. J., Deighton, M. H., Williams, S. R., Kindermann, M. and Duval, S. 2015. An inhibitor persistently decreased enteric methane emission from dairy cows with no negative effect on milk production. *Proc. Natl. Acad. Sci. U.S.A.* 112(34), 10663–8. doi:10.1073/pnas.1504124112.

Hünerberg, M., McGinn, S. M., Beauchemin, K. A., Okine, E. K., Harstad, O. M. and McAllister, T. A. 2013a. Effect of dried distillers grains plus solubles on enteric methane emissions and nitrogen excretion from growing beef cattle1. *J. Anim. Sci.* 91(6), 2846–57. doi:10.2527/jas.2012-5564.

Hünerberg, M., McGinn, S. M., Beauchemin, K. A., Okine, E. K., Harstad, O. M. and McAllister, T. A. 2013b. Effect of dried distillers' grains with solubles on enteric methane emissions and nitrogen excretion from finishing beef cattle. *Can. J. Anim. Sci.* 93(3), 373–85. doi:10.4141/cjas2012-151.

Hünerberg, M., Little, S. M., Beauchemin, K. A., McGinn, S. M., O'Connor, D., Okine, E. K., Harstad, O. M., Kröbel, R. and McAllister, T. A. 2014. Feeding high concentrations of corn dried distillers' grains decreases methane, but increases nitrous oxide emissions from beef cattle production. *Agric. Sys.* 127, 19–27. doi:10.1016/j.agsy.2014.01.005.

Huws, S. A., Creevey, C. J., Oyama, L. B., Mizrahi, I., Denman, S. E., Popova, M., Muñoz-Tamayo, R., Forano, E., Waters, S. M., Hess, M., Tapio, I., Smidt, H., Krizsan, S. J., Yáñez-Ruiz, D. R., Belanche, A., Guan, L., Gruninger, R. J., McAllister, T. A., Newbold, C. J., Roehe, R., Dewhurst, R. J., Snelling, T. J., Watson, M., Suen, G., Hart, E. H., Kingston-Smith, A. H., Scollan, N. D., do Prado, R. M., Pilau, E. J., Mantovani, H. C., Attwood, G. T., Edwards, J. E., McEwan, N. R., Morrisson, S., Mayorga, O. L., Elliott, C. and Morgavi, D. P. 2018. Addressing global ruminant agricultural challenges through understanding the rumen microbiome: past, present, and future. *Front. Microbiol.* 9, 2161. doi:10.3389/fmicb.2018.02161.

IPCC. 2006. *IPCC Guidelines for National Greenhouse Gas Inventories - A Primer.* Institute for Global Environmental Strategies, Japan.

Ishaq, S. L., AlZahal, O., Walker, N. and McBride, B. 2017. An investigation into rumen fungal and protozoal diversity in three rumen fractions, during high-fiber or grain-induced sub-acute ruminal acidosis conditions, with or without active dry yeast supplementation. *Front. Microbiol.* 8, 1943. doi:10.3389/fmicb.2017.01943.

Iwamoto, M., Asanuma, N. and Hino, T. 2002. Ability of *Selenomonas ruminantium*, *Veillonella parvula*, and *Wolinella succinogenes* to reduce nitrate and nitrite with special reference to the suppression of ruminal methanogenesis. *Anaerobe* 8(4), 209–15. doi:10.1006/anae.2002.0428.

Jia, X., Wang, M., Yuan, W., Shah, S., Shi, W., Meng, X., Ju, X. and Yang, B. 2016. N_2O emission and nitrogen transformation in chicken manure and biochar co-composting. *Trans. ASABE* 59(5), 1277–83. doi:10.13031/trans.59.11685.

Jin, D., Zhao, S., Zheng, N., Bu, D., Beckers, Y., Denman, S. E., McSweeney, C. S. and Wang, J. 2017. Differences in ureolytic bacterial composition between the rumen digesta and rumen wall based on ureC gene classification. *Front. Microbiol.* 8, 385–. doi:10.3389/fmicb.2017.00385.

Johnson, K. A. and Johnson, D. E. 1995. Methane emissions from cattle. *J. Anim. Sci.* 73(8), 2483-92. doi:10.2527/1995.7382483x.

Jordan, G., Predotova, M., Ingold, M., Goenster, S., Dietz, H., Joergensen, R. G. and Buerkert, A. 2015. Effects of activated charcoal and tannin added to compost and to soil on carbon dioxide, nitrous oxide and ammonia volatilization. *J. Plant Nutr. Soil Sci.* 178(2), 218-28. doi::doi:10.1002/jpln.201400233.

Joseph, S., Pow, D., Dawson, K., Mitchell, D. R. G., Rawal, A., Hook, J., Taherymoosavi, S., Van Zwieten, L., Rust, J., Donne, S., Munroe, P., Pace, B., Graber, E., Thomas, T., Nielsen, S., Ye, J., Lin, Y., Pan, G., Li, L. and Solaiman, Z. M. 2015. Feeding biochar to cows: an innovative solution for improving soil fertility and farm productivity. *Pedosphere* 25(5), 666-79. doi:10.1016/S1002-0160(15)30047-3.

Joseph, S., Kammann, C. I., Shepherd, J. G., Conte, P., Schmidt, H. P., Hagemann, N., Rich, A. M., Marjo, C. E., Allen, J., Munroe, P., Mitchell, D. R. G., Donne, S., Spokas, K. and Graber, E. R. 2018. Microstructural and associated chemical changes during the composting of a high temperature biochar: mechanisms for nitrate, phosphate and other nutrient retention and release. *Sci. Total Environ.* 618, 1210-23. doi:10.1016/j.scitotenv.2017.09.200.

Joye, S. B. 2012. Microbiology: a piece of the methane puzzle. *Nature* 491(7425), 538-9. doi:10.1038/nature11749.

Judy, J. V., Brown-Brandl, T. M., Fernando, S. C. and Kononoff, P. J. 2016. 1454 Manipulation of lactating dairy cows diets using reduced-fat distillers' grains, corn oil, and calcium sulfate to reduce methane production measured by indirect calorimetry. *J. Anim. Sci.* 94(suppl_5), 706-. doi:10.2527/jam2016-1454.

Kajikawa, H., Valdes, C., Hillman, K., Wallace, R. J. and J Newbold, C. 2003. Methane oxidation and its coupled electron-sink reactions in ruminal fluid. *Lett. Appl. Microbiol.* 36(6), 354-7. doi:10.1046/j.1472-765X.2003.01317.x.

Karhu, K., Mattila, T., Bergström, I. and Regina, K. 2011. Biochar addition to agricultural soil increased CH$_4$ uptake and water holding capacity - results from a short-term pilot field study. *Agric. Ecosyst. Environ.* 140(1-2), 309-13. doi:10.1016/j.agee.2010.12.005.

Kim, J. N., Méndez-García, C., Geier, R. R., Iakiviak, M., Chang, J., Cann, I. and Mackie, R. I. 2017. Metabolic networks for nitrogen utilization in Prevotella ruminicola 23. *Sci. Rep.* 7(1), 7851. doi:10.1038/s41598-017-08463-3.

Klieve, A. V., Ouwerkerk, D. and Maguire, A. J. 2012. Archaea in the foregut of macropod marsupials: PCR and amplicon sequence-based observations. *J. Appl. Microbiol.* 113(5), 1065-75. doi::doi:10.1111/j.1365-2672.2012.05428.x.

Knapp, J. R., Laur, G. L., Vadas, P. A., Weiss, W. P. and Tricarico, J. M. 2014. Invited review: enteric methane in dairy cattle production: quantifying the opportunities and impact of reducing emissions. *J. Dairy Sci.* 97(6), 3231-61. doi:10.3168/jds.2013-7234.

Koenig, K. M. and Beauchemin, K. A. 2018. Effect of feeding condensed tannins in high protein finishing diets containing corn distillers grains on ruminal fermentation, nutrient digestibility, and route of nitrogen excretion in beef cattle. *J. Anim. Sci.* 96(10), 4398-413. doi:10.1093/jas/sky273.

Koenig, K. M., Beauchemin, K. A. and McGinn, S. M. 2018. Feeding condensed tannins to mitigate ammonia emissions from beef feedlot cattle fed high-protein finishing diets containing distillers grains. *J. Anim. Sci.* 96(10), 4414-30. doi:10.1093/jas/sky274.

Koneswaran, G. and Nierenberg, D. 2008. Global farm animal production and global warming: impacting and mitigating climate change. *Environ. Health Perspect.* 116(5), 578-82. doi:10.1289/ehp.11034.

Kraus, T. E. C., Zasoski, R. J., Dahlgren, R. A., Horwath, W. R. and Preston, C. M. 2004. Carbon and nitrogen dynamics in a forest soil amended with purified tannins from different plant species. *Soil Biol. Biochem.* 36(2), 309-21. doi:10.1016/j.soilbio.2003.10.006.

Kumar, S., Indugu, N., Vecchiarelli, B. and Pitta, D. W. 2015. Associative patterns among anaerobic fungi, methanogenic archaea, and bacterial communities in response to changes in diet and age in the rumen of dairy cows. *Front. Microbiol.* 6, 781-. doi:10.3389/fmicb.2015.00781.

Lamar, R. T., Olk, D. C., Mayhew, L. and Bloom, P. R. 2014. A new standardized method for quantification of humic and fulvic acids in humic ores and commercial products. *J. AOAC Int.* 97(3), 721-30. doi:10.5740/jaoacint.13-393.

Latham, E. A., Anderson, R. C., Pinchak, W. E. and Nisbet, D. J. 2016. Insights on alterations to the rumen ecosystem by nitrate and nitrocompounds. *Front. Microbiol.* 7, 228. doi:10.3389/fmicb.2016.00228.

Lee, C. and Beauchemin, K. A. 2014. A review of feeding supplementary nitrate to ruminant animals: nitrate toxicity, methane emissions, and production performance. *Can. J. Anim. Sci.* 94(4), 557-70. doi:10.4141/cjas-2014-069.

Published by Burleigh Dodds Science Publishing Limited, 2020.

Lee, C., Araujo, R. C., Koenig, K. M. and Beauchemin, K. A. 2015a. Effects of encapsulated nitrate on eating behavior, rumen fermentation, and blood profile of beef heifers fed restrictively or ad libitum. *J. Anim. Sci.* 93(5), 2405–18. doi:10.2527/jas.2014-8851.

Lee, C., Araujo, R. C., Koenig, K. M. and Beauchemin, K. A. 2015b. Effects of encapsulated nitrate on enteric methane production and nitrogen and energy utilization in beef heifers. *J. Anim. Sci.* 93(5), 2391–404. doi:10.2527/jas.2014-8845.

Leng, R. A. 2014. Interactions between microbial consortia in biofilms: a paradigm shift in rumen microbial ecology and enteric methane mitigation. *Anim. Prod. Sci.* 54(5), 519–43. doi:10.1071/AN13381.

Leng, R., Preston, T. and Inthapanya, S. 2012. Biochar reduces enteric methane and improves growth and feed conversion in local "Yellow" cattle fed cassava root chips and fresh cassava foliage. *Livest. Res. Rural Dev.* 24. Available at: http://www.lrrd .org/lrrd24/11/leng24199.htm.

Li, L., Davis, J., Nolan, J. and Hegarty, R. 2012. An initial investigation on rumen fermentation pattern and methane emission of sheep offered diets containing urea or nitrate as the nitrogen source. *Anim. Prod. Sci.* 52(7), 653–8. doi:10.1071/AN11254.

Li, F., Wang, Z., Dong, C., Li, F., Wang, W., Yuan, Z., Mo, F. and Weng, X. 2017. Rumen bacteria communities and performances of fattening lambs with a lower or greater subacute ruminal acidosis risk. *Front. Microbiol.* 8, 2506. doi:10.3389/fmicb.2017.02506.

Liu, L., Xu, X., Cao, Y., Cai, C., Cui, H. and Yao, J. 2017. Nitrate decreases methane production also by increasing methane oxidation through stimulating NC10 population in ruminal culture. *AMB Express* 7(1), 76–. doi:10.1186/s13568-017-0377-2.

Lloyd, K. G., Lapham, L. and Teske, A. 2006. An anaerobic methane-oxidizing community of ANME-1b archaea in hypersaline Gulf of Mexico sediments. *Appl. Environ. Microbiol.* 72(11), 7218–30. doi:10.1128/AEM.00886-06.

Lopes, J. C., de Matos, L. F., Harper, M. T., Giallongo, F., Oh, J., Gruen, D., Ono, S., Kindermann, M., Duval, S. and Hristov, A. N. 2016. Effect of 3-nitrooxypropanol on methane and hydrogen emissions, methane isotopic signature, and ruminal fermentation in dairy cows. *J. Dairy Sci.* 99(7), 5335–44. doi:10.3168/jds.2015-10832.

Lyons, T., Boland, T., Storey, S. and Doyle, E. 2017. Linseed oil supplementation of lambs' diet in early life leads to persistent changes in rumen microbiome structure. *Front. Microbiol.* 8, 1656. doi:10.3389/fmicb.2017.01656.

Maeda, K., Hanajima, D., Toyoda, S., Yoshida, N., Morioka, R. and Osada, T. 2011. Microbiology of nitrogen cycle in animal manure compost. *Microb. Biotechnol.* 4(6), 700–9. doi:10.1111/j.1751-7915.2010.00236.x.

Maeda, K., Spor, A., Edel-Hermann, V., Heraud, C., Breuil, M. C., Bizouard, F., Toyoda, S., Yoshida, N., Steinberg, C. and Philippot, L. 2015. N$_2$O production, a widespread trait in fungi. *Sci. Rep.* 5, 9697–. doi:10.1038/srep09697.

Mao, S. Y., Huo, W. J. and Zhu, W. Y. 2016. Microbiome-metabolome analysis reveals unhealthy alterations in the composition and metabolism of ruminal microbiota with increasing dietary grain in a goat model. *Environ. Microbiol.* 18(2), 525–41. doi:10.1111/1462-2920.12724.

Martin, C., Ferlay, A., Mosoni, P., Rochette, Y., Chilliard, Y. and Doreau, M. 2016. Increasing linseed supply in dairy cow diets based on hay or corn silage: effect on enteric methane emission, rumen microbial fermentation, and digestion. *J. Dairy Sci.* 99(5), 3445–56. doi:10.3168/jds.2015-10110.

Martinez, C. M., Alvarez, L. H., Celis, L. B. and Cervantes, F. J. 2013. Humus-reducing microorganisms and their valuable contribution in environmental processes. *Appl. Microbiol. Biotechnol.* 97(24), 10293–308. doi:10.1007/s00253-013-5350-7.

Martínez-Fernández, G., Abecia, L., Arco, A., Cantalapiedra-Hijar, G., Martín-García, A. I., Molina-Alcaide, E., Kindermann, M., Duval, S. and Yáñez-Ruiz, D. R. 2014. Effects of ethyl-3-nitrooxy propionate and 3-nitrooxypropanol on ruminal fermentation, microbial abundance, and methane emissions in sheep. *J. Dairy Sci.* 97(6), 3790–9. doi:10.3168/jds.2013-7398.

Martinez-Fernandez, G., Denman, S. E., Yang, C., Cheung, J., Mitsumori, M. and McSweeney, C. S. 2016. Methane inhibition alters the microbial community, hydrogen flow, and fermentation response in the rumen of cattle. *Front. Microbiol.* 7, 1122–. doi:10.3389/fmicb.2016.01122.

Martinez-Fernandez, G., Denman, S. E., Cheung, J. and McSweeney, C. S. 2017. Phloroglucinol degradation in the rumen promotes the capture of excess hydrogen generated from methanogenesis inhibition. *Front. Microbiol.* 8, 1871. doi:10.3389/fmicb.2017.0187.

Martinez-Fernandez, G., Duval, S., Kindermann, M., Schirra, H. J., Denman, S. E. and McSweeney, C. S. 2018. 3-NOP vs. halogenated compound: methane production, ruminal fermentation and microbial community response in forage fed cattle. *Front. Microbiol.* 9, 1582. doi:10.3389/fmicb.2018.01582.

Masoom, H., Courtier-Murias, D., Farooq, H., Soong, R., Kelleher, B. P., Zhang, C., Maas, W. E., Fey, M., Kumar, R., Monette, M., Stronks, H. J., Simpson, M. J. and Simpson, A. J. 2016. Soil organic matter in its native state: unravelling the most complex biomaterial on earth. *Environ. Sci. Technol.* 50(4), 1670–80. doi:10.1021/acs.est.5b03410.

McAllister, T. A., Bae, H. D., Yanke, L. J., Cheng, K. J. and Muir, A. 1994. Effect of condensed tannins from birdsfoot trefoil on endoglucanase activity and the digestion of cellulose filter paper by ruminal fungi. *Can. J. Microbiol.* 40(4), 298–305. doi:10.1139/m94-048.

McGinn, S. M., Chung, Y.-H., Beauchemin, K. A., Iwaasa, A. D. and Grainger, C. 2009. Use of corn distillers' dried grains to reduce enteric methane loss from beef cattle. *Can. J. Anim. Sci.* 89(3), 409–13. doi:10.4141/CJAS08133.

McMurphy, C. P., Duff, G. C., Harris, M. A., Sanders, S. R., Chirase, N. K., Bailey, C. R. and Ibrahim, R. M. 2009. Effect of humic/fulvic acid in beef cattle finishing diets on animal performance, ruminal ammonia and serum urea nitrogen concentration. *J. Appl. Anim. Res.* 35(2), 97–100. doi:10.1080/09712119.2009.9706995.

McSweeney, C., Palmer, B., Kennedy, P. and Krause, D. 1998. Effect of *Calliandra tannins* on rumen microbial function. *Anim. Prod. Aust* 22, 289–.

McSweeney, C. S., Palmer, B., McNeill, D. M. and Krause, D. O. 2001. Microbial interactions with tannins: nutritional consequences for ruminants. *Anim. Feed Sci. Technol.* 91(1-2), 83–93. doi:10.1016/S0377-8401(01)00232-2.

Miller, K. E., Lai, C.-T., Friedman, E. S., Angenent, L. T. and Lipson, D. A. 2015. Methane suppression by iron and humic acids in soils of the Arctic Coastal Plain. *Soil Biol. Biochem.* 83, 176–83. doi:10.1016/j.soilbio.2015.01.022.

Mitsumori, M., Ajisaka, N., Tajima, K., Kajikawa, H. and Kurihara, M. 2002. Detection of Proteobacteria from the rumen by PCR using methanotroph-specific primers. *Lett. Appl. Microbiol.* 35(3), 251–5. doi::doi:10.1046/j.1472-765X.2002.01172.x.

Morgavi, D. P., Forano, E., Martin, C. and Newbold, C. J. 2010. Microbial ecosystem and methanogenesis in ruminants. *Animal* 4(7), 1024-36. doi:10.1017/S1751731110000546.

Muñoz, C., Hube, S., Morales, J. M., Yan, T. and Ungerfeld, E. M. 2015. Effects of concentrate supplementation on enteric methane emissions and milk production of grazing dairy cows. *Livest. Sci.* 175, 37-46. doi:10.1016/j.livsci.2015.02.001.

Mutabaruka, R., Hairiah, K. and Cadisch, G. 2007. Microbial degradation of hydrolysable and condensed tannin polyphenol-protein complexes in soils from different land-use histories. *Soil Biol. Biochem.* 39(7), 1479-92. doi:10.1016/j.soilbio.2006.12.036.

National Research Council. 2008. *Acute Exposure Guideline Levels for Selected Airborne Chemicals: Volume 6.* Available at: https://www.ncbi.nlm.nih.gov/books/NBK207883/.

Naumann, H. D., Tedeschi, L. O., Zeller, W. E. and Huntley, N. F. 2017. The role of condensed tannins in ruminant animal production: advances, limitations and future directions. *Rev. Bras. Zootec.* 46(12), 929-49. doi:10.1590/s1806-92902017001200009.

Niu, M., Appuhamy, J. A. D. R. N., Leytem, A. B., Dungan, R. S. and Kebreab, E. 2016. Effect of dietary crude protein and forage contents on enteric methane emissions and nitrogen excretion from dairy cows simultaneously. *Anim. Prod. Sci.* 56(3). doi:10.1071/AN15498.

Nolan, J. V., Hegarty, R. S., Hegarty, J., Godwin, I. R. and Woodgate, R. 2010. Effects of dietary nitrate on fermentation, methane production and digesta kinetics in sheep. *Anim. Prod. Sci.* 50(8), 801-6. doi:10.1071/AN09211.

Norton, J. M. 2008. Nitrification in agricultural soils. In: Schepers, J. S. and Raun, W. R. (Eds), *Nitrogen in Agricultural Systems.* American Society of Agronomy, Crop Science Society of America, Soil Science Society of America, Madison, WI, pp. 173-99.

Oldick, B. S. and Firkins, J. L. 2000. Effects of degree of fat saturation on fiber digestion and microbial protein synthesis when diets are fed twelve times daily. *J. Anim. Sci.* 78(9), 2412-20. doi:10.2527/2000.7892412x.

Olijhoek, D. W., Hellwing, A. L. F., Brask, M., Weisbjerg, M. R., Højberg, O., Larsen, M. K., Dijkstra, J., Erlandsen, E. J. and Lund, P. 2016. Effect of dietary nitrate level on enteric methane production, hydrogen emission, rumen fermentation, and nutrient digestibility in dairy cows. *J. Dairy Sci.* 99(8), 6191-205. doi:10.3168/jds.2015-10691.

Opio, C., Gerber, P., Mottet, A., Falcucci, A., Tempio, G., MacLeod, M., Vellinga, T., Henderson, B. and Steinfeld, H. 2013. *Greenhouse Gas Emissions from Ruminant Supply Chains–A Global Life Cycle Assessment.* Food and Agriculture Organization of the United Nations (FAO), Rome.

Parmar, N. R., Nirmal Kumar, J. I. and Joshi, C. G. 2015. Exploring diet-dependent shifts in methanogen and methanotroph diversity in the rumen of Mehsani buffalo by a metagenomics approach. *Front. Life Sci.* 8(4), 371-8. doi:10.1080/21553769.2015.1063550.

Patra, A. K. and Saxena, J. 2011. Exploitation of dietary tannins to improve rumen metabolism and ruminant nutrition. *J. Sci. Food Agric.* 91(1), 24-37. doi:10.1002/jsfa.4152.

Patra, A. K. and Yu, Z. 2013. Effects of coconut and fish oils on ruminal methanogenesis, fermentation, and abundance and diversity of microbial populations *in vitro*. *J. Dairy Sci.* 96(3), 1782-92. doi:10.3168/jds.2012-6159.

Patra, A. K. and Yu, Z. 2014. Combinations of nitrate, saponin, and sulfate additively reduce methane production by rumen cultures in vitro while not adversely affecting feed digestion, fermentation or microbial communities. *Bioresour. Technol.* 155, 129–35. doi:10.1016/j.biortech.2013.12.099.

Plaizier, J. C., Li, S., Tun, H. M. and Khafipour, E. 2017. Nutritional models of experimentally-induced subacute ruminal acidosis (SARA) differ in their impact on rumen and hindgut bacterial communities in dairy cows. *Front. Microbiol.* 7, 2128. doi:10.3389/fmicb.2016.02128.

Ponce, C. H., Arteaga, C. and Flores, A. 2016. Effects of humic acid supplementation on pig growth performance, nitrogen digestibility, odor, and ammonia emission. *J. Anim. Sci.* 94(suppl_5), 486. doi:10.2527/jam2016-1016.

Powell, J. M., Aguerre, M. J. and Wattiaux, M. A. 2011. Dietary crude protein and tannin impact dairy manure chemistry and ammonia emissions from incubated soils. *J. Environ. Qual.* 40(6), 1767–74. doi:10.2134/jeq2011.0085.

Ramos, A. F. O., Terry, S. A., Holman, D. B., Breves, G., Pereira, L. G. R., Silva, A. G. M. and Chaves, A. V. 2018. Tucumã oil shifted ruminal fermentation, reducing methane production and altering the microbiome but decreased substrate digestibility within a RUSITEC fed a mixed hay – concentrate diet. *Front. Microbiol.* 9, 1647–. doi:10.3389/fmicb.2018.01647.

Reynolds, C. K., Humphries, D. J., Kirton, P., Kindermann, M., Duval, S. and Steinberg, W. 2014. Effects of 3-nitrooxypropanol on methane emission, digestion, and energy and nitrogen balance of lactating dairy cows. *J. Dairy Sci.* 97(6), 3777–89. doi:10.3168/jds.2013-7397.

Richards, M. A., Lie, T. J., Zhang, J., Ragsdale, S. W., Leigh, J. A. and Price, N. D. 2016. Exploring hydrogenotrophic methanogenesis: a genome scale metabolic reconstruction of Methanococcus maripaludis. *J. Bacteriol.* 198, 3379–90. doi:10.1128/JB.00571-16.

Robertson, G. P. and Vitousek, P. M. 2009. Nitrogen in agriculture: balancing the cost of an essential resource. *Annu. Rev. Environ. Resour.* 34(1), 97–125. doi:10.1146/annurev.environ.032108.105046.

Romero-Perez, A., Okine, E. K., McGinn, S. M., Guan, L. L., Oba, M., Duval, S. M., Kindermann, M. and Beauchemin, K. A. 2014. The potential of 3-nitrooxypropanol to lower enteric methane emissions from beef cattle1. *J. Anim. Sci.* 92(10), 4682–93. doi:10.2527/jas.2014-7573.

Romero-Perez, A., Okine, E. K., McGinn, S. M., Guan, L. L., Oba, M., Duval, S. M., Kindermann, M. and Beauchemin, K. A. 2015. Sustained reduction in methane production from long-term addition of 3-nitrooxypropanol to a beef cattle diet. *J. Anim. Sci.* 93(4), 1780–91. doi:10.2527/jas.2014-8726.

Saleem, A. M., Ribeiro, G. O., Yang, W. Z., Ran, T., Beauchemin, K. A., McGeough, E. J., Ominski, K. H., Okine, E. K. and McAllister, T. A. 2018. Effect of engineered biocarbon on rumen fermentation, microbial protein synthesis, and methane production in an artificial rumen (RUSITEC) fed a high forage diet. *J. Anim. Sci.* 96(8), 3121–30. doi:10.1093/jas/sky204.

Sauvant, D., Giger-Reverdin, S., Serment, A. and Broudiscou, L. 2011. Influences des régimes et de leur fermentation dans le rumen sur la production de méthane par les ruminants. *INRA Product. Anim.* 24, 433–46.

Scheid, D., Stubner, S. and Conrad, R. 2003. Effects of nitrate- and sulfate-amendment on the methanogenic populations in rice root incubations. *FEMS Microbiol. Ecol.* 43(3), 309–15. doi:10.1111/j.1574-6941.2003.tb01071.x.

Shanks, O. C., Kelty, C. A., Archibeque, S., Jenkins, M., Newton, R. J., McLellan, S. L., Huse, S. M. and Sogin, M. L. 2011. Community structures of fecal bacteria in cattle from different animal feeding operations. *Appl. Environ. Microbiol.* 77(9), 2992–3001. doi:10.1128/AEM.02988-10.

Sharma, P., Laor, Y., Raviv, M., Medina, S., Saadi, I., Krasnovsky, A., Vager, M., Levy, G. J., Bar-Tal, A. and Borisover, M. 2017. Compositional characteristics of organic matter and its water-extractable components across a profile of organically managed soil. *Geoderma* 286, 73–82. doi:10.1016/j.geoderma.2016.10.014.

Sheng, P., Ribeiro, G. O., Wang, Y. and McAllister, T. A. 2017. Humic substances supplementation reduces ruminal methane production and increases the efficiency of microbial protein synthesis in vitro. *J. Anim. Sci.* 95(suppl_4), 300. doi:10.2527/asasann.2017.613.

Shi, Y., Parker, D., Cole, N., Auvermann, B. and Mehlhorn, J. 2001. Surface amendments to minimize ammonia emissions from beef cattle feedlots. *Transactions of the ASAE* 44(3), 677–82. doi:10.13031/2013.6105.

Steiner, C., Das, K. C., Melear, N. and Lakly, D. 2010. Reducing nitrogen loss during poultry litter composting using biochar. *J. Environ. Qual.* 39(4), 1236–42. doi:10.2134/jeq2009.0337.

Stevenson, F. J. 1995. Humus Chemistry: Genesis, Composition, Reactions, Second Edition. *J. Chem. Educ.* 72, A93. doi:10.1021/ed072pA93.6.

Tan, H. Y., Sieo, C. C., Lee, C. M., Abdullah, N., Liang, J. B. and Ho, Y. W. 2011. Diversity of bovine rumen methanogens *in vitro* in the presence of condensed tannins, as determined by sequence analysis of 16S rRNA gene library. *J. Microbiol.* 49(3), 492–8. doi:10.1007/s12275-011-0319-7.

Tan, W., Jia, Y., Huang, C., Zhang, H., Li, D., Zhao, X., Wang, G., Jiang, J. and Xi, B. 2018. Increased suppression of methane production by humic substances in response to warming in anoxic environments. *J. Environ. Manage.* 206, 602–6. doi:10.1016/j.jenvman.2017.11.012.

Tapio, I., Fischer, D., Blasco, L., Tapio, M., Wallace, R. J., Bayat, A. R., Ventto, L., Kahala, M., Negussie, E., Shingfield, K. J. and Vilkki, J. 2017a. Taxon abundance, diversity, co-occurrence and network analysis of the ruminal microbiota in response to dietary changes in dairy cows. *PLoS ONE* 12(7), e0180260. doi:10.1371/journal.pone.0180260.

Tapio, I., Snelling, T. J., Strozzi, F. and Wallace, R. J. 2017b. The ruminal microbiome associated with methane emissions from ruminant livestock. *J. Anim. Sci. Biotechnol.* 8, 7. doi:10.1186/s40104-017-0141-0.

Tawheed, M. E. S. and Baowei, Z. 2017. Review paper: the fundamentals of biochar as a soil amendment tool and management in agriculture scope: an overview for farmers and gardeners. *J. Agric. Chem. Environ.* 6, 38–61. doi:10.4236/jacen.2017.61003.

Terry, S. A., Ramos, A. F. O., Holman, D. B., McAllister, T. A., Breves, G. and Chaves, A. V. 2018a. Humic substances alter ammonia production and the microbial populations within a RUSITEC fed a mixed hay – concentrate diet. *Front. Microbiol.* 9, 1410. doi:10.3389/fmicb.2018.01410.

Terry, S. A., Ribeiro, G. O., Gruninger, R. J., Hunerberg, M., Ping, S., Chaves, A. V., Burlet, J., Beauchemin, K. A. and McAllister, T. A. 2018b. Effect of humic substances on rumen fermentation, nutrient digestibility, methane emissions, and rumen microbiota in beef heifers. *J. Anim. Sci.* 96, 3863–77. doi:10.1093/jas/sky265.

Terry, S. A., Ribeiro, G. O., Gruninger, R. J., Chaves, A. V., Beauchemin, K. A., Okine, E. and McAllister, T. A. 2019. A pine enhanced biochar does not decrease enteric CH4 emissions, but alters the rumen microbiota. *Front. Vet. Sci.* 6, 308. doi:10.3389/fvets.2019.00308.

Turk, J. 2016. Meeting projected food demands by 2050: understanding and enhancing the role of grazing ruminants. *J. Anim. Sci.* 94(suppl_6), 53–62. doi:10.2527/jas.2016-0547.

Ungerfeld, E. M. 2015a. Limits to dihydrogen incorporation into electron sinks alternative to methanogenesis in ruminal fermentation. *Front. Microbiol.* 6, 1272. doi:10.3389/fmicb.2015.01272.

Ungerfeld, E. M. 2015b. Shifts in metabolic hydrogen sinks in the methanogenesis-inhibited ruminal fermentation: a meta-analysis. *Front. Microbiol.* 6, 37–. doi:10.3389/fmicb.2015.00037.

Ungerfeld, E. and Kohn, R. 2006. The role of thermodynamics in the control of ruminal fermentation. In: Sejrsen, K., Hvelplund, T. and Nielsen, M. O. (Eds), *Ruminant Physiology: Digestion, Metabolism and Impact of Nutrition on Gene Expression, Immunology and Stress*. Wageningen Academic Publishers, Wageningen, Netherlands, pp. 55–85.

van Wyngaard, J. D. V., Meeske, R. and Erasmus, L. J. 2018. Effect of concentrate feeding level on methane emissions, production performance and rumen fermentation of Jersey cows grazing ryegrass pasture during spring. *Anim. Feed Sci. Technol.* 241, 121–32. doi:10.1016/j.anifeedsci.2018.04.025.

van Zijderveld, S. M., Gerrits, W. J. J., Apajalahti, J. A., Newbold, J. R., Dijkstra, J., Leng, R. A. and Perdok, H. B. 2010. Nitrate and sulfate: effective alternative hydrogen sinks for mitigation of ruminal methane production in sheep. *J. Dairy Sci.* 93(12), 5856–66. doi:10.3168/jds.2010-3281.

van Zijderveld, S. M., Gerrits, W. J. J., Dijkstra, J., Newbold, J. R., Hulshof, R. B. A. and Perdok, H. B. 2011. Persistency of methane mitigation by dietary nitrate supplementation in dairy cows. *J. Dairy Sci.* 94(8), 4028–38. doi:10.3168/jds.2011-4236.

Varadyova, Z., Kisidayová, S. and Jalc, D. 2009. Effect of humic acid on fermentation and ciliate protozoan population in rumen fluid of sheep *in vitro. J. Sci. Food Agric.* 89(11), 1936–41. doi:10.1002/jsfa.3675.

Vyas, D., McGinn, S. M., Duval, S. M., Kindermann, M. and Beauchemin, K. A. 2016. Effects of sustained reduction of enteric methane emissions with dietary supplementation of 3-nitrooxypropanol on growth performance of growing and finishing beef cattle1. *J. Anim. Sci.* 94(5), 2024–34. doi:10.2527/jas.2015-0268.

Vyas, D., Alemu, A. W., McGinn, S. M., Duval, S. M., Kindermann, M. and Beauchemin, K. A. 2018a. The combined effects of supplementing monensin and 3-nitrooxypropanol on methane emissions, growth rate, and feed conversion efficiency in beef cattle fed high-forage and high-grain diets. *J. Anim. Sci.* 96(7), 2923–38. doi:10.1093/jas/sky174.

Vyas, D., McGinn, S. M., Duval, S. M., Kindermann, M. K. and Beauchemin, K. A. 2018b. Optimal dose of 3-nitrooxypropanol for decreasing enteric methane emissions from beef cattle fed high-forage and high-grain diets. *Anim. Prod. Sci.* 58(6), 1049–55. doi:10.1071/AN15705.

Waghorn, G. 2008. Beneficial and detrimental effects of dietary condensed tannins for sustainable sheep and goat production–progress and challenges. *Anim. Feed Sci. Technol.* 147(1–3), 116–39. doi:10.1016/j.anifeedsci.2007.09.013.

Wallace, R. J., Rooke, J. A., McKain, N., Duthie, C. A., Hyslop, J. J., Ross, D. W., Waterhouse, A., Watson, M. and Roehe, R. 2015. The rumen microbial metagenome associated with high methane production in cattle. *BMC Genomics* 16, 839. doi:10.1186/s12864-015-2032-0.

Woodward, S. L., Waghorn, G. C. and Laboyrie, P. G. 2004. Condensed tannins in birdsfoot trefoil (*Lotus corniculatus*) reduce methane emissions from dairy cows. In: *Proceedings of the New Zealand Society of Animal Production*. New Zealand Society of Animal Production, Hamilton, New Zealand, pp. 160–4.

Yuan, Y., Chen, H., Yuan, W., Williams, D., Walker, J. T. and Shi, W. 2017. Is biochar-manure co-compost a better solution for soil health improvement and N_2O emissions mitigation? *Soil Biol. Biochem.* 113, 14–25. doi:10.1016/j.soilbio.2017.05.025.

Zhang, J., Shi, H., Wang, Y., Li, S., Cao, Z., Ji, S., He, Y. and Zhang, H. 2017. Effect of dietary forage to concentrate ratios on dynamic profile changes and interactions of ruminal microbiota and metabolites in Holstein heifers. *Front. Microbiol.* 8, 2206. doi:10.3389/fmicb.2017.02206.

Zhao, L., Meng, Q., Ren, L., Liu, W., Zhang, X., Huo, Y. and Zhou, Z. 2015. Effects of nitrate addition on rumen fermentation, bacterial biodiversity and abundance. *Asian-Australas. J. Anim. Sci.* 28(10), 1433–41. doi:10.5713/ajas.15.0091.

Zhao, L., Meng, Q., Li, Y., Wu, H., Huo, Y., Zhang, X. and Zhou, Z. 2018. Nitrate decreases ruminal methane production with slight changes to ruminal methanogen composition of nitrate-adapted steers. *BMC Microbiol.* 18(1), 21. doi:10.1186/s12866-018-1164-1.

Zhou, S., Xu, J., Yang, G. and Zhuang, L. 2014. Methanogenesis affected by the co-occurrence of iron(III) oxides and humic substances. *FEMS Microbiol. Ecol.* 88(1), 107–20. doi:10.1111/1574-6941.12274.

Zhu, Z., Kristensen, L., Difford, G. F., Poulsen, M., Noel, S. J., Abu Al-Soud, W., Sørensen, S. J., Lassen, J., Løvendahl, P. and Højberg, O. 2018. Changes in rumen bacterial and archaeal communities over the transition period in primiparous Holstein dairy cows. *J. Dairy Sci.* 101(11), 9847–62. doi:10.3168/jds.2017-14366.

Chapter 4

Host-rumen microbiome interactions and influences on feed conversion efficiency (FCE), methane production and other productivity traits

Elie Jami, Agricultural Research Organization – Volcani Center, Israel; and Itzhak Mizrahi, Ben-Gurion University of the Negev, Israel

1 Introduction

The rumen microbiome has the important task of supplying ruminants with most of their dietary requirements and is responsible for providing them with up to 70% of their metabolic needs and protein supply (Siciliano-Jones and Murphy, 1989; Bergman, 1990). This tremendous feat is possible due to the large diversity of microorganisms in the rumen, operating on different trophic levels (Flint et al., 2008; Moraïs and Mizrahi, 2019a,b). As such, ruminants are considered the hallmark of obligatory host-microbe interactions. The notion that differences in microbial composition could affect the animals' physiology, efficiency and waste output, has been suggested for the past 70 years of ruminant research, much earlier than the development of high throughput technology, which allowed researchers in the past decade to establish such a link (Krause et al., 2003). Hungate, in his seminal book *The Rumen and Its Microbes* (Hungate, 2013), suggested that a modulation of the microbial community toward improving fiber digestion may represent an avenue for increased productivity. Indeed the early works regarding the rumen microbiome largely focused on a subset of cultivable bacteria for which improved function was

http://dx.doi.org/10.19103/AS.2020.0067.18

thought to increase animal productivity (Krause et al., 2003). However, these early attempts at modifying the rumen microbial composition toward improved efficiency have mostly failed to sustain a desired phenotype (Attwood et al., 1988; Flint et al., 1989; Wallace and Walker, 1993; Miyagi et al., 1995; Krause et al., 1999, 2003). It has been pointed out that, in order to achieve this goal, one has to first understand the role of each component of the microbiome and its effect on the overall microbial community and the host.

Today, with our ability to assess the composition of the rumen microbial community as a whole, a new holistic view of the microbiome has emerged, whereby application of basic ecological principles on the overall microbiome structure and the physiological response of the host can be studied. This will lead us to an increased understanding of the role of the microbiome and its components on production efficiency, health and waste emissions such as methane. This chapter focuses on the recent discovery about the role of the ruminant microbiome on energy harvest, methane emission and the potential genetic factors determining its microbial composition and selection.

2 Core community, resilience and natural variation in rumen microbiome composition

2.1 Core community

Identifying common microbial features can lead to an understanding of the more basic requirements of the rumen ecosystem as they likely serve key functions in rumen metabolism. Several independent studies recognize the existence of a core microbial community in the rumen shared between and within ruminant lineages (Jami and Mizrahi, 2012a; Henderson et al., 2015). A comprehensive analysis of the microbiome of 32 ruminant and pseudo-ruminant species (Henderson et al., 2015) emphasized the shared and divergent nature of the rumen microbiome composition across a wide geographical range, animal lineages and management conditions (Henderson et al., 2015). The authors identified a core community of taxa at the genus and species level, shared across different ruminant animal lineages. These include *Prevotella,* the most dominant genus in the rumen, *Butyrivibrio* and *Ruminococcus*, which harbor the main cellulolytic species in the rumen, as well as unclassified Lachnospiraceae, Ruminococcaceae, Clostridiales (all Firmicutes) and Bacteroidales (Bacteroidetes) for bacteria. The bovine rumen was also shown to be particularly enriched with Fibrobacter, an important cellulolytic species, shown to be most abundant in cattle-fed high-forage diet (Henderson et al., 2015). The members of the *Methanobrevibacter gottschalkii* and *Methanobrevibacter ruminantium* clades represent the core methanogenic community observed in the rumen (Henderson et al., 2015). This shared presence of bacterial taxa across different foregut animal lineages suggests that the core taxa have a key role in rumen metabolism and function (Shade and Handelsman, 2012). This suggestion is reinforced by the recent observation that species defined as core are more associated with physiological traits of the host in dairy cattle (Li et al., 2019; Wallace et al., 2019).

2.2 Variation

In all model animals tested, including ruminants, inter-individual variation in both microbial composition and abundance exists between animals despite stringent account for external factors such as housing, management and diet (Brulc et al., 2009; Jami and Mizrahi, 2012a; Henderson et al., 2015). Brulc et al. (2009) conducted a pioneering study characterizing the rumen microbial composition using shotgun metagenomics sequencing and showed broad differences in the community composition of three steers. One of the steers exhibited a different composition compared with the other two steers despite being under the same diet. Similarly, the characterization of 16 dairy cows under the same diet and management condition exhibited a 0.51 average pairwise similarity using the Bray-Curtis index, which takes into account the presence, and abundance of taxa. Some genera appeared to be relatively stable in terms of presence and abundance while others can exhibit up to two orders of magnitude differences in abundance between the cows (Jami and Mizrahi, 2012a,b).

2.3 Resilience

While studies have consistently pointed out that samples taken from different host animals can exhibit high variation in composition (Brulc et al., 2009; Li et al., 2009; Jami and Mizrahi, 2012a), microbiome assessment across different sampling time points within the same cow reveals a remarkable stability (Li et al., 2009; Welkie et al., 2009). In a study using automated ribosomal intergenic spacer analysis (ARISA) to examine the changes in ruminal bacterial communities during the feeding cycle, similar observations were made, emphasizing both the stability of the rumen microbial community when established within a cow and the large differences in composition between different cows (Welkie et al., 2009). In steers, long-term temporal assessment of the changes in microbiome composition after dietary change revealed that, after 25 days on a new diet, the microbiome shows little variation hereafter (Snelling et al., 2019). The microbiome is resilient to such an extent that even large perturbations, such as transfaunation, where the rumen fluid of one cow is almost completely replaced with the rumen fluid of another, showed that within just a few weeks, rumen microbiome content reverted to a composition more closely resembling the original (Weimer et al., 2017). This stresses that host factors may strongly influence microbial assembly. Recent studies show a connection between the individual animals' genetics and its respective microbiome, as well as heritability of some rumen microbiome components (Roehe et al., 2016; Li et al., 2016; Sasson et al., 2017; Wallace et al., 2019). A larger experiment showed that following transfaunation, each individual cow exhibited unique patterns of reestablishment further strengthening the possibility of large host effect on the microbial community (Zhou et al., 2018).

3 Microbiome-dependent traits

The fermentation products of rumen microbial activity serve as the main source of energy for the animal, contributing to up to 70% of their metabolic requirements (Wolin, 1979; Bergman, 1990). Plant feed ingested undergoes a cascade of degradation, from complex polymers to intermediate molecules, reaching end products which are either absorbed by the animal or emitted to the environment. Several methods and indexes were developed to account for the level of efficiency in which feed is being converted into usable products by the animal for growth and production, including feed conversion ratio (FCR), residual feed intake (RFI), energy corrected milk/dry matter intake (ECM/DMI), and the more recent residual intake and gain (RGI) (NRC, 2001; Berry and Crowley, 2012). Although each of these indexes calculate efficiency differently and encompass different observed physiological parameters of the ruminant, the overall rationale behind them remains similar in their attempt to measure the overall ratio between the energetic value of the feed or diet ingested compared with the energy absorbed by the ruminant for maintenance and production (Mizrahi, 2012). These global traits are supplemented by more specific assessment, depending on the research question, such as protein, carbohydrate, and lipid content in milk (Jami et al., 2014), health parameters of the host (Jewell et al., 2015), or rumen-specific metabolites linked to energy uptake such as VFA composition and quantification, or energy loss, such as methane (Hernandez-Sanabria et al., 2010; Shabat et al., 2016; Tapio et al., 2017). The proposed effects of microbiome composition and microbial gene expression on animal physiology, with emphasis on performance and methane emission, have been the focus of many studies in the past decade, summarized in Table 1. As mentioned above, the fundamental microbial functions of the plant fiber degradation and fermentation processes are similar between microbiomes across host animals (Moraïs and Mizrahi, 2019a). However, the variation in composition and abundance of specific microbial taxa and their gene expression were linked to methane emissions as well as specific rumen metabolites which have considerable impact on host traits such as milk composition and energy-harvest efficiency from the feed (Shi et al., 2014; Shabat et al., 2016; Kamke et al., 2016; Li and Guan, 2017). In cattle, species of the *Prevotella* genus have been implicated in both increased and decreased productivity and milk composition parameters (Carberry et al., 2012; Jami et al.,

Table 1 Summary list of studies assessing the link between rumen microbiome composition and animal physiology

Animal	Method	Main findings	Reference
Beef cattle	PCR-DGGE	Feed efficiency phenotype and specific bacteria linked to SCFA profile	Guan et al. (2008)
Beef cattle	PCR-DGGE	Differential prevalence of *M. stadtmanae* and specific *Methanobrevibacter* strains between high and low RFI. Overall higher methanogens taxonomic diversity in high RFI animals	Zhou et al. (2009)
Beef cattle	PCR-DGGE, qPCR	Specific bacterial and archaeal OTUs associated with specific VFAs and with different RFI phenotypes under low-energy diet	Hernandez-Sanabria et al. (2010)
Beef cattle	PCR-DGGE, qPCR	*Succinovibrio spp.* associated with low-methane emission. Acetate higher in high-efficient animals	Hernandez-Sanabria et al. (2012)
Beef cattle	PCR-DGGE and qPCR	High *Prevotella* abundance in low-efficiency animals. Link between bacterial profile and feed efficiency is different between diets.	Carberry et al. (2012)
Beef cattle	Clone library and 16S amplicon sequencing	*Methanobrevibacter* species differ in abundance between high and low RFI animals	Carberry et al. (2014)
Sheep	Shotgun metagenomics and metatranscriptomics	Higher gene expression related to the hydrogenotrophic methanogenesis pathway in high methane-emitting sheep	Shi et al. (2014)
Dairy cattle	16S amplicon sequencing	Correlations between bacterial genera and production parameters. *Prevotella* negatively correlates with milk fat yield	Jami et al. (2014)
Beef cattle	16S amplicon sequencing	High abundance of *Prevotella* linked to inefficient phenotype	McCann et al. (2014)
Sheep	16S amplicon sequencing	Identification of three different 'ruminotypes' associated with high- and low-methane emission	Kittelmann et al. (2014)

(Continued)

Table 1 (*Continued*)

Animal	Method	Main findings	Reference
Dairy cattle	16S amplicon sequencing	Core OTUs associated with either efficient (*Prevotella spp.*) or inefficient cows (*Prevotella* genus, *Butyrivibrio*)	Jewell et al. (2015)
Dairy cattle	16S Amplicon sequencing, shotgun metagenomics, metabolomics	*Megasphaera elsdenii* and the acrylate pathway involved in increased efficiency to high efficiency. Efficient cows exhibit lower bacterial diversity	Shabat et al. (2016)
Sheep	Metatranscriptomics	Low methane emission phenotype in sheep enriched in *Sharpea azabuensis* and *Megasphaera spp.* and the acrylate pathway	Kamke et al. (2016)
Dairy cattle	16S amplicon sequencing	*Prevotella*, S24-7 and Succinivibrionaceae lineages positively correlated with milk yield	Indugu et al. (2017)
Dairy cattle	16S amplicon sequencing	*M. ruminantium* and *M. gottschalkii* associated with low methane emission	Danielsson et al. (2017)
Beef cattle	Metatranscriptomics	Low feed-efficiency animals express a higher diversity of gene pathways	Li and Guan (2017)
Dairy cattle	Shotgun metagenomics	Higher abundance of Bacteroidetes and lower abundance of methanogens in efficient cows	Delgado et al. (2019)
Beef cattle	Shotgun metagenomics	Microbial genes related to cell wall biosynthesis, hemicellulose and cellulose degradation host-microbiome cross talk associated with FCR. Vitamin B12 biosynthesis, environmental information processing and bacterial mobility genes associated with RFI	Lima et al. (2019)
Dairy cattle	16S amplicon sequencing, metabolomics, host genotyping	Core microbes more associated and predictive of a wide range of host traits, including feed efficiency, methane emission and VFAs	Wallace et al. (2019)

DGGE = denaturing gradient gel electrophoresis; SCFA = short-chain fatty acids; RFI = residual feed intake; qPCR = quantitative PCR; OTUs = operational taxonomic units; VFA = volatile fatty acids; FCR = feed conversion ratio.

2014; McCann et al., 2014; Jewell et al., 2015; Shabat et al., 2016; Indugu et al., 2017). In a study assessing the correlation between taxa abundance of the microbiome components and physiological parameters of 15 cows, a positive correlation was observed between the ratio of Firmicutes/Bacteroidetes and daily fat production in the milk (Jami et al., 2014). The difference in ratio was mostly driven by the vast difference in the abundance of the *Prevotella* genus (Bacteroidetes), which negatively correlated with milk fat yield (Jami et al., 2014). These findings can be mirrored to observations from mice and human microbiome studies, in which a lower abundance of Bacteroidetes correlated with increased blood and tissue adiposity in mice (Turnbaugh et al., 2006). *Prevotella* and its species were also found to be linked to feed efficiency in a study investigating the dynamics of the microbial population in the cow's rumen throughout two lactation periods (Jewell et al., 2015). This study also found a negative correlation between specific operational taxonomic units (OTUs; at 97% similarity defined as species) associated with the *Prevotella* genus and production efficiency. However, the same study also identified various *Prevotella* species associated with higher feed efficiency (Jewell et al., 2015), along with another study showing that *Prevotella* was linked to increased milk production yield in dairy cattle (Indugu et al., 2017). It is likely that different *Prevotella* species affect rumen physiology differently as this genus is highly diverse in the rumen (Ley, 2016). Our functional understanding of this genus is restricted to a number of cultivated species, limiting our understanding of their full functional scope. The recent availability of a compendium of ~5000 genomes from the rumen using deep metagenomic sequencing may shed light on the diverse functions carried by the species belonging to the *Prevotella* genus (Stewart et al., 2019). In a series of studies performed on steers grouped according to their RFI values (low/high RFI), the bacterial and archaeal rumen community was assessed using PCR-DGGE fingerprinting. The authors showed that specific bacterial taxa had a higher likelihood of being present depending on the respective groups along with VFA production patterns (Guan et al., 2008; Hernandez-Sanabria et al., 2010). Furthermore, species diversity was lower in the low RFI steers, when methanogenic populations were compared using 16s rDNA clone libraries, and a correlation between the composition of methanogens and host efficiency could be observed (Zhou and Hernandez-Sanabria, 2009; Zhou et al., 2010). In a subsequent study, the ruminal methanogenic and bacterial populations of 58 steers which differed in feed efficiency and diet were analyzed (Hernandez-Sanabria et al., 2012). The authors found that *Succinivibrio* spp. and *Eubacterium* spp. were correlated with increased efficiency in steers. The authors suggested that the higher abundance of *Succinivibrio* in low RFI steers, along with the higher acetate and lower isovalerate production, shifts rumen metabolism away from methanogenesis and toward propionate production (Hernandez-Sanabria et al., 2012). Interestingly, the family Succinivibrionaceae and its species have been recurrently implicated in a low methane phenotype in sheep and tammar wallabies (Pope et al., 2011; Wallace et al., 2015b). A microbiome-dependent steering of metabolism away from methanogenesis and toward VFA production was also demonstrated in dairy cattle (Shabat et al., 2016). A study was conducted examining 78 cows with extreme RFI phenotype (38 inefficient, 40 efficient), bacterial taxonomic composition, bacterial gene composition and ecological features, which were significantly correlated and predictive of the feed-efficiency phenotype (Shabat et al., 2016). The authors showed that the microbiome of efficient cows had a lower diversity than that of inefficient cows, and a gene composition steered toward production of usable end products for the cows such as propionate. The study identified *Megasphaera elsdenii* and *Coprococcus catus*, and the acrylate pathway encoded by these species, as a central pathway linked to higher efficiency. The acrylate pathway uses the intermediate product lactate to produce propionate. This study proposed a model whereby an interplay exists between the acrylate pathway and VFA production and methanogenesis, suggesting that both phenotypes, though not necessarily converging, link to feed efficiency. These findings may further relate to the previously mentioned observations of increased Succinivibrionaceae taxa associated with a different VFA profile, lower methane emission and higher efficiency (Wallace et al., 2015b). *Succinivibrio dextrinosolvens*, a known resident of the rumen, is shown to increase its lactate production under specific growth conditions (O'Herrin and Kenealy, 1993). Integrated together, these results point toward a specific cascade of fermentation, with lactate as intermediate, being central in determining efficiency in ruminants. A caveat to these observation remains that link between physiological parameters and the microbiome might be diet specific and that different links can be observed across different diets (Carberry et al., 2012).

4 Methane production

Methane production is exclusively performed by methanogenic archaea (methanogens) inhabiting the rumen. Methanogens serve as electron sinks for the entire rumen ecosystem driving the directionality of the fermentation process, which would be otherwise inhibited by the H_2 produced (McAllister and Newbold, 2008; van Lingen et al., 2016). This is mostly performed by the hydrogenotrophic methanogenesis pathway, the main methanogenesis pathway in the rumen, in which CO_2 and H_2 products of fermentation are converted to methane. Although necessary, methane production constitutes an energy loss for the animal, ranging between 2 and 12%, in addition to the detrimental effect of methane on the environment (Johnson and Johnson, 1995). Thus, studies on the microbiome effect on methane production are intimately connected to those assessing its effect on production efficiency (Shabat et al., 2016). While being an integral part of the rumen ecosystem, ruminants can exhibit a wide variation in methane emission. Several studies have shown that higher methane emission is weak, or not at all linked to the absolute abundance of methanogens present in the rumen of cattle and sheep (Shi et al., 2014; Kittelmann et al., 2014; Tapio et al., 2017). However, a recent study using whole-genome sequencing of the rumen of steer found that the ratio of archaea bacteria is predictive of methane emission with a correlation of 0.49 (Wallace et al., 2015a).

Differential composition of methanogens and methanogenic gene expression has been more consistently implicated in the different methane emission phenotypes observed in cattle and sheep, and several methanogenic taxa have been linked to increased methane emission. It is generally agreed upon that the most abundant methanogenic genus in the mature rumen is the *Methanobrevibacter* (Janssen and Kirs, 2008; Friedman et al., 2017; Tapio et al., 2017). Several studies categorized this genus into two clades – SMGT (*smithii-gottschalkii-millerae-thaurei*) and RO (*ruminantium-olleyae*) (King et al., 2011) – with the high presence of SMGT clade being associated with higher methanogenesis potential (Danielsson et al., 2012, 2017; Shi et al., 2014). Using metatranscriptomics, Shi et al. (2014) observed that *Methanobrevibacter gottschalkii* abundance is increased in high methane-yielding sheep. The authors also noted that methanogenic transcript abundance of genes encoding to CO_2/H_2 pathway was significantly increased in high methane-emitting sheep while no difference was observed in terms of methanogen abundance.

Although methanogens are sole producers of methane, they rely on upstream fermentation processes and outputs by other microbial taxa, which were shown to affect its production. These include fermentation by bacteria, protozoa and fungi. Kittelmann et al. (2014) identified three different bacterial assemblies, termed 'ruminotypes', linked to different methane production phenotypes in sheep (Kittelmann et al., 2014). Ruminotypes associated with high methane emission exhibited an enrichment in H_2-producing bacteria, such as Ruminococcaceae, Lachnospiraceae, Catabacteriaceae, *Coprococcus*, other Clostridiales, *Prevotella*, Bacteroidales and Alphaproteobacteria. The ruminotypes associated with low methane emission was associated with either the propionate-producing *Quinella ovalis*, or with lactate and succinate producers such as *Fibrobacter* spp., *Kandleria vitulina*, *Olsenella* spp., *Prevotella bryantii* and *Sharpea azabuensis* (Kittelmann et al., 2014). As mentioned in the context of feed efficiency, lactate and succinate producers such as *Succinivibrio dextrinosolvens* have been suggested as being responsible for the low methane production observed in tammar wallabies and associated with high-feed efficiency in cattle (Pope et al., 2011; Wallace et al., 2015b). Another study similarly identified the lactate producer, *Sharpeae albenenzis*, enriched in the rumen of low methane-emitting sheep (Kamke et al., 2016). The author suggested that lactate production by this species diverts electrons in the form of H_2 from methanogenesis toward production of lactate, which in turn is converted to propionate through the acrylate pathway of *Megasphaera elsdenii*. These findings mirror the observed higher abundance of lactate-utilizing bacteria and genes observed in high efficiency, low CH_4-emitting dairy cows (Shabat et al., 2016). The overall aggregated conclusions of these findings suggest that differences in electron transfer between microbial species across the rumen trophic network result in differences in community states (Moraïs and Mizrahi, 2019a), leading to different outcomes in terms of output end products. The alternative community state model suggests that, for yet unresolved reasons, one rumen ecosystem may favor a metabolism steered toward H_2 production leading to higher methane production while another would favor lactate production and utilization toward VFA production (Moraïs and Mizrahi, 2019a).

In addition to the observed link between the bacterial domain and methanogenesis, rumen ciliate protozoa have been attributed many roles in enhancing methane emission through mutualistic associations with rumen methanogens (Newbold et al., 2015). This research topic has been the focus of many studies in the past three decades. As opposed to bacteria, ciliate protozoa are not essential to the proper functioning of the rumen, giving the opportunity to assess protozoa effect on the rumen ecosystem and host physiology using defaunation – the removal of protozoa from the rumen by various means. A recent meta-analyses summarizing 30 years of experimental defaunation studies presented significant physiological differences between faunated and defaunated ruminants (sheep and cattle), including protein availability to methane emission (Guyader et al., 2014; Newbold et al., 2015). Specifically, the meta-analyses revealed that defaunation consistently leads to a significant decrease in methane emission of up to 11% from the animal in vivo. In vitro studies further suggested that within the protozoa community, holotrichous protozoa are more involved in methane production (Belanche et al., 2015). Furthermore, an additional meta-analysis which compared methane emission to the protozoa cell abundance in 79 studies showed a strong linear correlation between protozoa cell numbers in vivo and methane emission (Guyader et al., 2014). As most protozoa in the rumen are known to produce high quantities of H_2, it has been suggested that the nature of this mutualism stems from the rich H_2 environment provided by protozoa favored by hydrogenotrophic methanogens (Newbold et al., 2015). Additional evidence of such a syntrophic mechanism was observed in a co-cultivation experiment between the protozoa *Polyplastron multivesiculatum* and the methanogen *Methanosarcina barkeri*, in which a decrease of H_2 in the culture coincided with an increase in methane (Ushida et al., 1997).

Using in situ hybridization, studies have shown that protozoa are largely colonized by methanogens, thereby strengthening the hypothesis of a strong mutualistic relationship between protozoa and methanogens (Finlay et al., 1994; Lloyd et al., 1996). Lloyd et al. (1996) showed that methanogens can be found attached to the outer pellicle of protozoa as well as inside the protozoa, but outside of the food vacuole, suggesting that they are not serving as food to the predatory protozoa, but rather as symbionts. Furthermore, protozoa were shown to carry a higher methanogens/bacteria ratio than the free-living fraction of the rumen, suggesting a specific tropism between methanogens and protozoa (Levy and Jami, 2018). However, when correcting for protozoa size, Belanche et al. (2014) did not observe such enrichment; therefore the question remains as to whether methanogens accumulate in a higher proportion than bacteria in and around protozoa (Belanche et al., 2014). While this question remains open, several studies have shown that the composition of methanogens associated with protozoa differs when compared to the free-living prokaryotic population (Tymensen et al., 2012; Tymensen and McAllister, 2012; Belanche et al., 2014; Levy and Jami, 2018). Taxa belonging to the *Methanobrevibacter* genus were overrepresented in the protozoa-associated fractions (Görtz, 2006; Tymensen et al., 2012; Belanche et al., 2014; Levy and Jami, 2018). This would support the notion that the methanogens-protozoa relationship is based on interspecies electron transfer via H_2, as species from the *Methanobrevibacter* genus are commonly associated with *hydrogenotrophic* methanogenesis (Janssen and Kirs, 2008). Furthermore, following the observation that no difference in methanogens abundance could be observed following defaunation, it was hypothesized that the methanogens associated with protozoa are more active in terms of methane production. Levy and Jami (2018) showed that, when separated by size, large protozoa exhibited a higher relative abundance of OTUs associated with the SGMT group of *Methanobrevibacter*, previously linked to high methane emission in cattle (Danielsson et al., 2017), supporting this hypothesis.

5 Nitrogen compounds: utilization and emission

Nitrogen cycling in the rumen ecosystem could potentially affect community composition and host attributes. In a study linking feed efficiency to the rumen microbiome and its functional capacities, it was shown that inefficient cows were enriched in functions related to protein digestion and amino acid biosynthesis (Shabat et al., 2016). Additionally, ciliate protozoa were shown to negatively affect N availability to the animal through increased predation of bacteria (Newbold et al., 2015). Specifically, *Entodinium* spp. were shown to decrease microbial protein availability to the host animal, and their elimination through defaunation was shown to increase protein supply by 30% (Newbold et al., 2015).

Proteolytic activity of microbes in the rumen can vary greatly between animals, with excess proteolytic activity, such as deamination, regarded as detrimental to efficient N utilization (Hartinger et al., 2018). Inefficient N utilization thus carries a negative effect on host production as well as the environment, contributing a large proportion of anthropogenic nitrous oxide released into the atmosphere (Huws et al., 2018). The differential abundance of specific community members such as hyper-ammonia-producing bacteria (HAB) may result in different utilization efficiency of N, and therefore control over such populations could improve nitrogen utilization by the animal (Firkins et al., 2007; Hartinger et al., 2018). This topic represents an important avenue for future studies in order to better understand the potential connection between the bacterial community composition and N cycling efficiency.

6 Microbiome and host genetics

While important strides were made in understanding the link between the microbiome and host physiology and productivity, the issue regarding the potential control of the host genetics on microbiome composition and selection remains vastly understudied, particularly compared with the plethora of studies on the topic in human microbiome research (Rothschild et al., 2018). The question whether host genetics affect the microbiome composition and subsequently microbial features related to energy harvest and methane emission, can have vast impact on our ability to rationally select genotypes in order to obtain favored phenotypes through selective breeding (Myer, 2019). In humans and mice, where arguably more data is available on the topic, several studies reported that the microbial community composition is partly dictated by the genetic makeup of the host (Benson et al., 2010; Goodrich et al., 2014, 2016; Bonder et al., 2016; Turpin et al., 2016). In contrast, a recent human study put into question the degree of influence that host genetics has on microbial composition and abundance (Rothschild et al., 2018). A first indication for a host-genetic influence on the composition of the microbial community and physiological parameters was derived by comparing the microbial composition of different cattle breeds and their hybrids (Guan et al., 2008; Hernandez-Sanabria et al., 2013; Paz et al., 2016). One such study performed on two different sire breeds showed a link between host genetics, microbial composition, specifically archaea:bacteria ratio, and methane, which would allow selection of animals based on genetics (Roehe et al., 2016). The authors proposed several mechanisms by which the host animal could control microbiome composition and subsequently its output, such as pH buffering by saliva production, or feed retention time in the rumen, the latter shown to be a heritable trait (Roehe et al., 2016). By combining microbial abundance data with the genomic profile of 47 dairy cows differing in efficiency, Sasson et al.

(2017) identified 22 OTUs associated with rumen-metabolic traits and host-physiological traits, which showed measurable heritability (Sasson et al., 2017). This study also shows that these heritable OTUs were more closely connected to host physiology and rumen metabolites than other rumen microbes. Recently, three large-scale studies examining hundreds of individual animals for establishing the connection of the host genetics to microbial composition using genome wide association (GWAS) approaches shared several similarities in their results (Difford et al., 2018; Li et al., 2019; Wallace et al., 2019). Difford et al. (2018), using a cohort of 750 dairy cattle, identified that the abundance of 6% of the bacterial community and 12% of archaeal community at the species level were heritable ($h^2 > 0.15$), that is, controlled by host genome, and that methane emission was associated with both microbiome features and host genetics, although both associations were largely independent (Difford et al., 2018). The independent nature of host genetic influence on methane emission and the microbiome suggests methane variation between animals is likely not a result of host genetics on the microbiome. The authors thus suggested that two parallel avenues should be considered in order to decrease methane emission: one related to breeding selection of low methane-associated traits within the animal genome and the second related to microbiome modulation toward mitigating methane emission (Difford et al., 2018). In contrast, in a cohort of 669 steers, a study showed that animal genetics contributes to the abundance of 59 microbial taxa, 56 bacteria and 3 archaea, which were also associated with host-feed efficiency traits and rumen-metabolic output (Li et al., 2019). The authors revealed 19 single nucleotide polymorphisms (SNPs), five of which are located in loci linked to feed efficiency, associated with 12 microbial taxa. Thus, this study suggests that host control of a subset of microbes in the rumen may have a direct effect on production efficiency. Furthermore, four of the observed heritable taxa were shown, using interaction network inference, to interact with a large number of taxa. In a large-scale GWAS performed throughout Europe, 1000 dairy cattle across four European countries were genotyped and the rumen microbiome composition was sequenced along with a large range of metabolic and physiological parameters. This study showed that taxa considered 'core', that is, recurrent in at least 50% of the cows within a specific farm, were connected to variation in host genetics, whereby their abundance can be explained to a significant extent by host genetics (Wallace et al., 2019). Within this core microbiome, 39 taxa were found to have measured heritability of up to $h^2 = 0.6$. Furthermore, using network inference against phenotypic traits, this study showed that those heritable core microbes are more linked to productivity parameters than to the non-core taxa. Moreover, in agreement with the study by Li et al. (2019), this study showed that these heritable microbes were central to the microbiome interaction networks, suggesting them as keystone species. These findings suggest that by controlling a subset of microbes, the host could have a larger indirect control over a broader range of microbes and therefore on rumen physiology. These observations validate the notion that the core microbial community is selected, and carry a large proportional role in defining animals' physiological traits. Although seemingly contradictory to the latest research on humans, some similarities in the results can be seen, while the differences in experimental design may explain the discrepancies. Firstly, as within humans, only a relatively small subset of taxa could be associated with host genetics (1.9%) (Rothschild et al., 2018), which is also true for cattle (0.25% of the OTUs) (Wallace et al., 2019). However, in terms of abundance, these taxa represented up to 60% of the microbiome in cattle while in humans the heritable taxa represent around 6% of the microbiome. Furthermore, animal studies exert a tighter control over many physiological and management parameters, as opposed to human studies, in which differences in diet between people for instance cannot be absolutely ascertained and controlled. Achieving homogeneity in environmental factors may indeed reveal the stronger impact of host genetics in determining microbiome structure. It is however clear that environmental factors such as diet likely exert the largest effect on microbiome compositions, but the heritability of output parameters such as methane emission suggests that host genetics may influence the fate of the alternative stable community states in each individual animal. This would lead to the possibility of devising novel criteria for selection based on the genetics of the ruminant in order to steer the microbiome toward an agriculturally favorable phenotype.

7 References

Attwood, G. T., Lockington, R. A., Xue, G. P. and Brooker, J. D. 1988. Use of a unique gene sequence as a probe to enumerate a strain of *Bacteroides ruminicola* introduced into the rumen. *Applied and Environmental Microbiology* 54, 534-9.

Belanche, A., de la Fuente, G. and Newbold, C. J. 2014. Study of methanogen communities associated with different rumen protozoal populations. *FEMS Microbiology Ecology* 90, 663-77.

Belanche, A., de la Fuente, G. and Newbold, C. J. 2015. Effect of progressive inoculation of fauna-free sheep with holotrich protozoa and total-fauna on rumen fermentation, microbial diversity and methane emissions. *FEMS Microbiology Ecology* 91(3).

Benson, A. K., Kelly, S. A., Legge, R., Ma, F., Low, S. J., Kim, J., Zhang, M., Oh, P. L., Nehrenberg, D., Hua, K., Kachman, S. D., Moriyama, E. N., Walter, J., Peterson, D. A. and Pomp, D. 2010. Individuality in gut microbiota composition is a complex polygenic trait shaped by multiple environmental and host genetic factors. *Proceedings of the National Academy of Sciences of the United States of America* 107, 18933-8.

Bergman, E. N. 1990. Energy contributions of volatile fatty acids from the gastrointestinal tract in various species. *Physiological Reviews* 70, 567-90.

Berry, D. P. and Crowley, J. J. 2012. Residual intake and body weight gain: a new measure of efficiency in growing cattle. *Journal of Animal Science* 90, 109-15.

Bonder, M. J., Kurilshikov, A., Tigchelaar, E. F., Mujagic, Z., Imhann, F., Vila, A. V., Deelen, P., Vatanen, T., Schirmer, M., Smeekens, S. P., Zhernakova, D. V., Jankipersadsing, S. A., Jaeger, M., Oosting, M., Cenit, M. C., Masclee, A. A. M., Swertz, M. A., Li, Y., Kumar, V., Joosten, L., Harmsen, H., Weersma, R. K., Franke, L., Hofker, M. H., Xavier, R. J., Jonkers, D., Netea, M. G., Wijmenga, C., Fu, J. and Zhernakova, A. 2016. The effect of host genetics on the gut microbiome. *Nature Genetics* 48, 1407–12.

Brulc, J. M., Antonopoulos, D. A., Miller, M. E., Wilson, M. K., Yannarell, A. C., Dinsdale, E. A., Edwards, R. E., Frank, E. D., Emerson, J. B., Wacklin, P., Coutinho, P. M., Henrissat, B., Nelson, K. E. and White, B. A. 2009. Gene-centric metagenomics of the fiber-adherent bovine rumen microbiome reveals forage specific glycoside hydrolases. *Proceedings of the National Academy of Sciences of the United States of America* 106, 1948–53.

Carberry, C. A., Kenny, D. A., Han, S., McCabe, M. S. and Waters, S. M. 2012. Effect of phenotypic residual feed intake and dietary forage content on the rumen microbial community of beef cattle. *Applied and Environmental Microbiology* 78, 4949–58.

Carberry, C. A., Waters, S. M., Waters, S. M., Kenny, D. A. and Creevey, C. J. 2014. Rumen methanogenic genotypes differ in abundance according to host residual feed intake phenotype and diet type. *Applied and Environmental Microbiology* 80, 586–94.

Danielsson, R., Schnürer, A. and Arthurson, V. 2012. Methanogenic population and CH_4 production in Swedish dairy cows fed different levels of forage. *Applied and Environmental Microbiology* 78(17), 6172–9.

Danielsson, R., Dicksved, J., Sun, L., Gonda, H., Müller, B., Schnürer, A. and Bertilsson, J. 2017. Methane production in dairy cows correlates with rumen methanogenic and bacterial community structure. *Frontiers in Microbiology* 8, 226.

Delgado, B., Bach, A., Guasch, I., González, C., Elcoso, G., Pryce, J. E. and Gonzalez-Recio, O. 2019. Whole rumen metagenome sequencing allows classifying and predicting feed efficiency and intake levels in cattle. *Scientific Reports* 9, 11.

Difford, G. F., Plichta, D. R., Løvendahl, P., Lassen, J., Noel, S. J., Højberg, O., Wright, A-D. G., Zhu, Z., Kristensen, L., Nielsen, H. B., Guldbrandtsen, B. and Sahana, G. 2018. Host genetics and the rumen microbiome jointly associate with methane emissions in dairy cows. *PLoS Genetics* 14, e1007580.

Finlay, B. J., Esteban, G., Clarke, K. J., Williams, A. G., Embley, T. M. and Hirt, R. P. 1994. Some rumen ciliates have endosymbiotic methanogens. *FEMS Microbiology Letters* 117, 157–61.

Firkins, J. L., Yu, Z. and Morrison, M. 2007. Ruminal nitrogen metabolism: perspectives for integration of microbiology and nutrition for dairy. *Journal of Dairy Science* 90 Suppl. 1, E1–16.

Flint, H. J., Bisset, J. and Webb, J. 1989. Use of antibiotic resistance mutations to track strains of obligately anaerobic bacteria introduced into the rumen of sheep. *The Journal of Applied Bacteriology* 67, 177–83.

Flint, H. J., Bayer, E. A., Rincon, M. T., Lamed, R. and White, B. A. 2008. Polysaccharide utilization by gut bacteria: potential for new insights from genomic analysis. *Nature Reviews. Microbiology* 6, 121–31.

Friedman, N., Jami, E. and Mizrahi, I. 2017. Compositional and functional dynamics of the bovine rumen methanogenic community across different developmental stages. *Environmental Microbiology* 19(8), 3365–73.

Goodrich, J. K., Waters, J. L., Poole, A. C., Sutter, J. L., Koren, O., Blekhman, R., Beaumont, M., Van Treuren, W., Knight, R., Bell, J. T., Spector, T. D., Clark, A. G. and Ley, R. E. 2014. Human genetics shape the gut microbiome. *Cell* 159, 789–99.

Goodrich, J. K., Davenport, E. R., Beaumont, M., Jackson, M. A., Knight, R., Ober, C., Spector, T. D., Bell, J. T., Clark, A. G. and Ley, R. E. 2016. Genetic determinants of the gut microbiome in UK twins. *Cell Host & Microbe* 19, 731-43.

Görtz, H.-D. 2006. Symbiotic associations between ciliates and prokaryotes. In: Dworkin, M., Falkow, S., Rosenberg, E., Schleifer, K.-H. and Stackebrandt, E. (Eds), *The Prokaryotes: Volume 1: Symbiotic Associations, Biotechnology, Applied Microbiology.* Springer New York, New York, NY, pp. 364-402.

Guan, L. L., Nkrumah, J. D., Basarab, J. A. and Moore, S. S. 2008. Linkage of microbial ecology to phenotype: correlation of rumen microbial ecology to cattle's feed efficiency. *FEMS Microbiology Letters* 288, 85-91.

Guyader, J., Eugène, M., Nozière, P., Morgavi, D. P., Doreau, M. and Martin, C. 2014. Influence of rumen protozoa on methane emission in ruminants: a meta-analysis approach. *Animal* 8, 1816-25.

Hartinger, T., Gresner, N. and Südekum, K.-H. 2018. Does intra-ruminal nitrogen recycling waste valuable resources? A review of major players and their manipulation. *Journal of Animal Science and Biotechnology* 9, 33.

Henderson, G., Cox, F., Ganesh, S., Jonker, A., Young, W., Global Rumen Census Collaborators and Janssen, P. H. 2015. Rumen microbial community composition varies with diet and host, but a core microbiome is found across a wide geographical range. *Scientific Reports* 5.

Hernandez-Sanabria, E., Guan, L. L., Goonewardene, L. A., Li, M., Mujibi, D. F., Stothard, P., Moore, S. S. and Leon-Quintero, M. C. 2010. Correlation of particular bacterial PCR-denaturing gradient gel electrophoresis patterns with bovine ruminal fermentation parameters and feed efficiency traits. *Applied and Environmental Microbiology* 76, 6338-50.

Hernandez-Sanabria, E., Goonewardene, L. A., Wang, Z., Durunna, O. N., Moore, S. S. and Guan, L. L. 2012. Impact of feed efficiency and diet on adaptive variations in the bacterial community in the rumen fluid of cattle. *Applied and Environmental Microbiology* 78, 1203-14.

Hernandez-Sanabria, E., Goonewardene, L. A., Wang, Z., Zhou, M., Moore, S. S. and Guan, L. L. 2013. Influence of sire breed on the interplay among rumen microbial populations inhabiting the rumen liquid of the progeny in beef cattle. *PLoS ONE* 8, e58461.

Hungate, R. E. 2013. *The Rumen and Its Microbes.* Elsevier.

Huws, S. A., Creevey, C. J., Oyama, L. B., Mizrahi, I., Denman, S. E., Popova, M., Muñoz-Tamayo, R., Forano, E., Waters, S. M., Hess, M., Tapio, I., Smidt, H., Krizsan, S. J., Yáñez-Ruiz, D. R., Belanche, A., Guan, L., Gruninger, R. J., McAllister, T. A., Newbold, C. J., Roehe, R., Dewhurst, R. J., Snelling, T. J., Watson, M., Suen, G., Hart, E. H., Kingston-Smith, A. H., Scollan, N. D., do Prado, R. M., Pilau, E. J., Mantovani, H. C., Attwood, G. T., Edwards, J. E., McEwan, N. R., Morrisson, S., Mayorga, O. L., Elliott, C. and Morgavi, D. P. 2018. Addressing global ruminant agricultural challenges through understanding the rumen microbiome: past, present, and future. *Frontiers in Microbiology* 9, 2161.

Indugu, N., Vecchiarelli, B., Baker, L. D., Ferguson, J. D., Vanamala, J. K. P. and Pitta, D. W. 2017. Comparison of rumen bacterial communities in dairy herds of different production. *BMC Microbiology* 17, 190.

Jami, E. and Mizrahi, I. 2012a. Composition and similarity of bovine rumen microbiota across individual animals. *PLoS ONE* 7, e33306.

Jami, E. and Mizrahi, I. 2012b. Similarity of the ruminal bacteria across individual lactating cows. *Anaerobe* 18, 338–43.

Jami, E., White, B. A. and Mizrahi, I. 2014. Potential role of the bovine rumen microbiome in modulating milk composition and feed efficiency. *PLoS ONE* 9, e85423.

Janssen, P. H. and Kirs, M. 2008. Structure of the archaeal community of the rumen. *Applied and Environmental Microbiology* 74, 3619–25.

Jewell, K. A., McCormick, C. A., Odt, C. L., Weimer, P. J. and Suen, G. 2015. Ruminal bacterial community composition in dairy cows is dynamic over the course of two lactations and correlates with feed efficiency. *Applied and Environmental Microbiology* 81, 4697–710.

Johnson, K. A. and Johnson, D. E. 1995. Methane emissions from cattle. *Journal of Animal Science* 73, 2483–92.

Kamke, J., Kittelmann, S., Soni, P., Li, Y., Tavendale, M., Ganesh, S., Janssen, P. H., Shi, W., Froula, J., Rubin, E. M. and Attwood, G. T. 2016. Rumen metagenome and metatranscriptome analyses of low methane yield sheep reveals a Sharpea-enriched microbiome characterised by lactic acid formation and utilisation. *Microbiome* 4, 56.

King, E. E., Smith, R. P., St-Pierre, B. and Wright, A.-D. G. 2011. Differences in the rumen methanogen populations of lactating Jersey and Holstein dairy cows under the same diet regimen. *Applied and Environmental Microbiology* 77, 5682–7.

Kittelmann, S., Pinares-Patiño, C. S., Seedorf, H., Kirk, M. R., Ganesh, S., McEwan, J. C. and Janssen, P. H. 2014. Two different bacterial community types are linked with the low-methane emission trait in sheep. *PLoS ONE* 9, e103171.

Krause, D. O., Smith, W. J., Ryan, F. M., Mackie, R. I. and McSweeney, C. S. 1999. Use of 16S-rRNA based techniques to investigate the ecological succession of microbial populations in the immature lamb rumen: tracking of a specific strain of inoculated ruminococcus and interactions with other microbial populations *in vivo*. *Microbial Ecology* 38, 365–76.

Krause, D. O., Denman, S. E., Mackie, R. I., Morrison, M., Rae, A. L., Attwood, G. T. and McSweeney, C. S. 2003. Opportunities to improve fiber degradation in the rumen: microbiology, ecology, and genomics. *FEMS Microbiology Reviews* 27, 663–93.

Levy, B. and Jami, E. 2018. Exploring the prokaryotic community associated within the rumen ciliate protozoa population. *Frontiers in Microbiology* 9, 2526.

Ley, R. E. 2016. Gut microbiota in 2015: prevotella in the gut: choose carefully. *Nature Reviews. Gastroenterology and Hepatology* 13, 69–70.

Li, F. and Guan, L. L. 2017. Metatranscriptomic profiling reveals linkages between the active rumen microbiome and feed efficiency in beef cattle. *Applied and Environmental Microbiology* 83(9).

Li, M., Penner, G. B., Hernandez-Sanabria, E., Oba, M. and Guan, L. L. 2009. Effects of sampling location and time, and host animal on assessment of bacterial diversity and fermentation parameters in the bovine rumen. *Journal of Applied Microbiology* 107, 1924–34.

Li, Z., Wright, A.-D. G., Si, H., Wang, X., Qian, W., Zhang, Z. and Li, G. 2016. Changes in the rumen microbiome and metabolites reveal the effect of host genetics on hybrid crosses. *Environmental Microbiology Reports* 8, 1016–23.

Li, F., Li, C., Chen, Y., Liu, J., Zhang, C., Irving, B., Fitzsimmons, C., Plastow, G. and Guan, L. L. 2019. Host genetics influence the rumen microbiota and heritable rumen microbial features associate with feed efficiency in cattle. *Microbiome* 7, 92.

Lima, J., Auffret, M. D., Stewart, R. D., Dewhurst, R. J., Duthie, C.-A., Snelling, T. J., Walker, A. W., Freeman, T. C., Watson, M. and Roehe, R. 2019. Identification of rumen microbial genes involved in pathways linked to appetite, growth, and feed conversion efficiency in cattle. *Frontiers in Genetics* 10, 701.

Lloyd, D., Williams, A. G., Amann, R., Hayes, A. J., Durrant, L. and Ralphs, J. R. 1996. Intracellular prokaryotes in rumen ciliate protozoa: detection by confocal laser scanning microscopy after in situ hybridization with fluorescent 16S rRNA probes. *European Journal of Protistology* 32, 523–31.

McAllister, T. A. and Newbold, C. J. 2008. Redirecting rumen fermentation to reduce methanogenesis. *Australian Journal of Experimental Agriculture* 48, 7–13.

McCann, J. C., Wiley, L. M., Forbes, T. D., Rouquette Jr., F. M. and, Tedeschi, L. O. 2014. Relationship between the rumen microbiome and residual feed intake-efficiency of brahman bulls stocked on bermudagrass pastures. *PLoS ONE* 9, e91864.

Miyagi, T., Kaneichi, K., Aminov, R. I., Kobayashi, Y., Sakka, K., Hoshino, S. and Ohmiya, K. 1995. Enumeration of transconjugated *Ruminococcus albus* and its survival in the goat rumen microcosm. *Applied and Environmental Microbiology* 61, 2030–2.

Mizrahi, I. 2012. The role of the rumen microbiota in determining the feed efficiency of dairy cows. In: *Beneficial Microorganisms in Multicellular Life Forms*, pp. 203–10. Springer.

Moraïs, S. and Mizrahi, I. 2019a. The road not taken: the rumen microbiome, functional groups, and community states. *Trends in Microbiology* 27, 538–49.

Moraïs, S. and Mizrahi, I. 2019b. Islands in the stream: from individual to communal fiber degradation in the rumen ecosystem. *FEMS Microbiology Reviews* 43, 362–79.

Myer, P. R. 2019. Bovine genome-microbiome interactions: metagenomic frontier for the selection of efficient productivity in cattle systems. *mSystems* 4(3).

National Research Council. 2001. *Nutrient Requirements of Dairy Cattle*. National Academy Press, Washington DC.

Newbold, C. J., de la, Fuente, G., Belanche, A., Ramos-Morales, E. and McEwan, N. R. 2015. The role of ciliate protozoa in the rumen. *Frontiers in Microbiology* 6.

O'Herrin, S. M. and Kenealy, W. R. 1993. Glucose and carbon dioxide metabolism by *Succinivibrio dextrinosolvens*. *Applied and Environmental Microbiology* 59, 748–55.

Paz, H. A., Anderson, C. L., Muller, M. J., Kononoff, P. J. and Fernando, S. C. 2016. Rumen bacterial community composition in holstein and jersey cows is different under same dietary condition and is not affected by sampling method. *Frontiers in Microbiology* 7, 1206.

Pope, P. B., Smith, W., Denman, S. E., Tringe, S. G., Barry, K., Hugenholtz, P., McSweeney, C. S., McHardy, A. C. and Morrison, M. 2011. Isolation of *Succinivibrionaceae* implicated in low methane emissions from tammar wallabies. *Science* 333, 646–8.

Roehe, R., Dewhurst, RJ, Duthie, C-A, Rooke, JA, McKain, N., Ross, DW, Hyslop, JJ, Waterhouse, A., Freeman, T. C., Watson, M. and Wallace, R. J. 2016. Bovine host genetic variation influences rumen microbial methane production with best selection criterion for low methane emitting and efficiently feed converting hosts based on metagenomic gene abundance. *PLoS Genetics* 12, e1005846.

Rothschild, D., Weissbrod, O., Barkan, E., Kurilshikov, A., Korem, T., Zeevi, D., Costea, P. I., Godneva, A., Kalka, I. N., Bar, N., Shilo, S., Lador, D., Vila, A. V., Zmora, N., Pevsner-Fischer, M., Israeli, D., Kosower, N., Malka, G., Wolf, B. C., Avnit-Sagi, T., Lotan-Pompan, M., Weinberger, A., Halpern, Z., Carmi, S., Fu, J., Wijmenga, C., Zhernakova, A., Elinav,

E. and Segal, E. 2018. Environment dominates over host genetics in shaping human gut microbiota. *Nature* 555, 210–15.

Sasson, G., Kruger, B-S. S., Seroussi, E., Doron-Faigenboim, A., Shterzer, N., Yaacoby, S., Berg, M. M. E., White, B. A., Halperin, E. and Mizrahi, I. 2017. Heritable bovine rumen bacteria are phylogenetically related and correlated with the cow's capacity to harvest energy from its feed. *mBio* 8.

Shabat, S. K. B., Sasson, G., Doron-Faigenboim, A., Durman, T., Yaacoby, S., Berg, Miller, M. E., White, B. A., Shterzer, N. and Mizrahi, I. 2016. Specific microbiome-dependent mechanisms underlie the energy harvest efficiency of ruminants. *The ISME Journal* 10, 2958.

Shade, A. and Handelsman, J. 2012. Beyond the Venn diagram: the hunt for a core microbiome. *Environmental Microbiology* 14, 4–12.

Shi, W., Moon, C. D., Leahy, S. C., Kang, D., Froula, J., Kittelmann, S., Fan, C., Deutsch, S., Gagic, D. and Seedorf, H. 2014. Methane yield phenotypes linked to differential gene expression in the sheep rumen microbiome. *Genome Research* 24, 1517–25.

Siciliano-Jones, J. and Murphy, M. R. 1989. Production of volatile fatty acids in the rumen and cecum-colon of steers as affected by forage: concentrate and forage physical form. *Journal of Dairy Science* 72, 485–92.

Snelling, T. J., Auffret, M. D., Duthie, C-A, Stewart, R. D., Watson, M., Dewhurst, R. J., Roehe, R. and Walker, A. W. 2019. Temporal stability of the rumen microbiota in beef cattle, and response to diet and supplements. *Animal Microbiome* 1, 16.

Stewart, R. D., Auffret, M. D., Warr, A., Walker, A. W., Roehe, R. and Watson, M. 2019. Compendium of 4,941 rumen metagenome-assembled genomes for rumen microbiome biology and enzyme discovery. *Nature Biotechnology* 37, 953–61.

Tapio, I., Snelling, T. J., Strozzi, F. and Wallace, R. J. 2017. The ruminal microbiome associated with methane emissions from ruminant livestock. *Journal of Animal Science and Biotechnology* 8, 7.

Turnbaugh, P. J., Ley, R. E., Mahowald, M. A., Magrini, V., Mardis, E. R. and Gordon, J. I. 2006. An obesity-associated gut microbiome with increased capacity for energy harvest. *Nature* 444, 1027–31.

Turpin, W., Espin-Garcia, O., Xu, W., Silverberg, M. S., Kevans, D., Smith, M. I., Guttman, D. S., Griffiths, A., Panaccione, R., Otley, A., Xu, L., Shestopaloff, K., Moreno-Hagelsieb G, G. E. M. Project Research Consortium, Paterson, A. D. and Croitoru, K. 2016. Association of host genome with intestinal microbial composition in a large healthy cohort. *Nature Genetics* 48, 1413–17.

Tymensen, L. D. and McAllister, T. A. 2012. Community structure analysis of methanogens associated with rumen protozoa reveals bias in universal archaeal primers. *Applied and Environmental Microbiology* 78, 4051–6.

Tymensen, L. D., Beauchemin, K. A. and McAllister, T. A. 2012. Structures of free-living and protozoa-associated methanogen communities in the bovine rumen differ according to comparative analysis of 16S rRNA and mcrA genes. *Microbiology* 158, 1808–17.

Ushida, K., Newbold, C. J. and Jouany, J.-P. 1997. Interspecies hydrogen transfer between the rumen ciliate *Polyplastron multivesiculatum* and, Methanosarcina barkeri. *The Journal of General and Applied Microbiology* 43, 129–31.

van Lingen, H. J., Plugge, C. M., Fadel, J. G., Kebreab, E., Bannink, A. and Dijkstra, J. 2016. Thermodynamic driving force of hydrogen on rumen microbial metabolism: a theoretical investigation. *PLoS ONE* 11, e0161362.

Wallace, R. J. and Walker, N. D. 1993. Isolation and attempted introduction of sugar alcohol-utilizing bacteria in the sheep rumen. *The Journal of Applied Bacteriology* 74, 353–9.

Wallace, R. J., John, Wallace, R., Rooke, J. A., Duthie, C-A, Hyslop, J. J., Ross, D. W., McKain, N., de, Souza, S. M., Snelling, T. J., Waterhouse, A. and Roehe, R. 2015a. Archaeal abundance in post-mortem ruminal digesta may help predict methane emissions from beef cattle. *Scientific Reports* 4.

Wallace, R. J., Rooke, J. A., McKain, N., Duthie, C.-A., Hyslop, J. J., Ross, D. W., Waterhouse, A., Watson, M. and Roehe, R. 2015b. The rumen microbial metagenome associated with high methane production in cattle. *BMC Genomics* 16, 839.

Wallace, R. J., Sasson, G., Garnsworthy, P. C., Tapio, I., Gregson, E., Bani, P., Huhtanen, P., Bayat, A. R., Strozzi, F., Biscarini, F., Snelling, T. J., Saunders, N., Potterton, S. L., Craigon, J., Minuti, A., Trevisi, E., Callegari, M. L., Cappelli, F. P., Cabezas-Garcia, E. H., Vilkki, J., Pinares-Patino, C., Fliegerová K. O., Mrázek, J., Sechovcová H, Kopečný J, Bonin, A., Boyer, F., Taberlet, P., Kokou, F., Halperin, E., Williams, J. L., Shingfield, K. J. and Mizrahi, I. 2019. A heritable subset of the core rumen microbiome dictates dairy cow productivity and emissions. *Science Advances* 5, eaav8391.

Weimer, P. J., Cox, M. S., de, Paula, T. V., Lin, M., Hall, M. B. and Suen, G. 2017. Transient changes in milk production efficiency and bacterial community composition resulting from near-total exchange of ruminal contents between high-and low-efficiency Holstein cows. *Journal of Dairy Science* 100, 7165–82.

Welkie, D. G., Stevenson, D. M. and Weimer, P. J. 2009. ARISA analysis of ruminal bacterial community dynamics in lactating dairy cows during the feeding cycle. *Anaerobe* 16, 94–100.

Wolin, M. J. 1979. The rumen fermentation: a model for microbial interactions in anaerobic ecosystems. In: Alexander, M. (Ed.), *Advances in Microbial Ecology* (vol. 3). Springer US, Boston, MA, pp. 49–77.

Zhou, M. I. and Hernandez-Sanabria, E. 2009. Assessment of the microbial ecology of ruminal methanogens in cattle with different feed efficiencies. *Applied and Environmental Microbiology* 75, 6524–33.

Zhou, M., Hernandez Sanabria, E. and Guan, L. 2009. Assessment of the microbial ecology of ruminal methanogens in cattle with different feed efficiencies. *Applied and Environmental Microbiology* 75, 6524–33. doi:10.1128/AEM.02815-081.

Zhou, M., Hernandez-Sanabria, E. and Guan, L. L. 2010. Characterization of variation in rumen methanogenic communities under different dietary and host feed efficiency conditions, as determined by PCR-denaturing gradient gel electrophoresis analysis. *Applied and Environmental Microbiology* 76, 3776–86.

Zhou, M., Peng, Y.-J., Chen, Y., Klinger, C. M., Oba, M., Liu, J.-X. and Guan, L. L. 2018. Assessment of microbiome changes after rumen transfaunation: implications on improving feed efficiency in beef cattle. *Microbiome* 6, 62.

Chapter 5

Developing closed-loop dairy value chains and tools to support decision-makers

Jack B. Hetherington, University of Adelaide/CSIRO Agriculture and Food/Fight Food Waste Cooperative Research Centre, Australia; Pablo Juliano, CSIRO Agriculture and Food, Australia; and Rodolfo García-Flores, CSIRO Data61, Australia

1 Introduction

The dairy sector is an important contributor to economies and diets globally, employing many people in the production of food that provides an important source of protein and micro-nutrients to consumers. However, dairy systems also contribute to environmental problems through emissions of greenhouse gases (GHG) (i.e. methane (CH_4) and nitrous oxide (N_2O)) and ammonia (NH_3) throughout the process of cattle digestion and manure management (Gac et al., 2007; Knapp et al., 2014). Additionally, losses and waste of dairy products along the value chain present a twofold problem. The first problem is the direct effects of decomposing organic material in landfill, which contributes to GHG emissions (FAO, 2013), and the discharge of pollutants into the environment (Smithers, 2008). The second problem is the inefficient use of resources such as water, fertiliser, electricity and fuel that have gone into producing, processing, cooling and transporting the food product until the point of disposal (Kummu et al., 2012).

http://dx.doi.org/10.19103/AS.2023.0120.22

Reduction of food loss and waste (FLW) has been recognised in the United Nation's (UN) Sustainable Development Goals as a global challenge for food systems, with significant economic, social and environmental ramifications. Kummu et al. (2012) estimated that approximately one-quarter of global water, fertiliser and cropland used to produce food are wasted. FLW also contributes significantly to GHG and nitrogen emissions (FAO, 2013) causing environmental issues such as contamination of freshwater, ocean acidification and damage to biosphere integrity (Lade et al., 2020). The social implications of FLW include lost meals that could feed food-insecure people (Reynolds et al., 2015). In the case of dairy products, this means loss of macro-nutrients (e.g. proteins) and micro-nutrients (e.g. calcium, choline, riboflavin, zinc and vitamin B12) (Chen et al., 2020; de Wit, 2001). FLW also adds significant costs to households, businesses and society more broadly. A study by Hanson and Mitchell (2017) of over 700 businesses across 17 countries found that, for every dollar invested in addressing FLW, there was a net return of $14. This is because properly managing FLW can reduce input, labour and overhead costs and even generate new revenue through developing new products or selling co-products as an input to other industries.

FLW is generated at different stages of the value chain and varies significantly across regions (Gustavsson et al., 2011). As shown in Fig. 1, the majority of dairy FLW in developing countries is generated during production, storage, distribution and processing, whereas in developed countries, most of the waste occurs at the consumption stage of the chain. There are many opportunities to utilise these waste streams and by-products in close-looped systems or the circular economy. This has been defined as 'an economic system that replaces the "end-of-life" concept with reducing, alternatively reusing, recycling and recovering materials in production/distribution and consumption processes' (Kirchherr et al., 2017, p. 229).

This chapter focuses on issues and opportunities for managing dairy FLW for countries with developed dairy value chains, i.e. economies with established

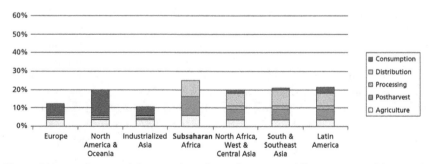

Figure 1 Loss and waste of dairy products for each region at different stages of the supply chain. Source: Reproduced with permission from: Gustavsson et al. (2011).

refrigeration infrastructure, traceability and quality control. While there will be some relevance to regions with less developed supply chains, this chapter does not intend to address the full spectrum of issues affecting these regions (Lore et al., 2005).

In countries with developed dairy chains, like Australia, dairy products generate different types of 'avoidable' and 'unavoidable' FLW, which refers to the components that are intended for human consumption (Papargyropoulou et al., 2014). Examples of avoidable dairy FLW include liquid milk or cheese products that have expired (i.e. reached their 'use by date') and are no longer safe to consume. A key example of often unavoidable dairy FLW is the generation of whey as a by-product of small-scale cheese production. As discussed at length in this book chapter, both avoidable and unavoidable FLW can be used as a valuable resource, such as an input to another business or transformed into new food products. Therefore, they do not always have to be considered waste.

In Australia, FIAL (2021) estimated that most of the waste from liquid milk products occurs within households, while most of the waste generated from cheese products occurs at the processing stage (see Fig. 2). The FIAL (2021) study did not consider any dairy by-product fed to animals as this was considered a desirable management option and therefore not measured as 'waste'. In the case of cheese whey, an estimated 3.1 million tonnes were produced in Australia in the 2018-9 financial year, with two-thirds being used as animal feed (personal communication, PA Bontinck, Lifecycles, 9 December 2021). This highlights there is significantly more whey that needs to be managed. Additionally, it underscores that the study did not consider the standard definition for FLW and its measurement, which is described in the FLW Protocol (Hanson et al., 2016). Furthermore, the whey accounted into animal feed ignores that there are high-value return options for its use. For example,

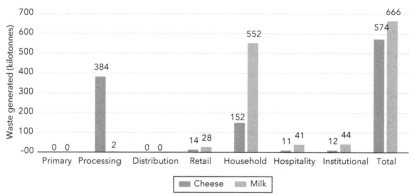

Figure 2 Quantity of dairy product waste along the value chain in Australia in 2018-9. Source: personal communication, PA Bontinck, Lifecycles, 9 December 2021.

instead of feeding whey to pigs, it may be more valuable to manufacture it into new food products or other high-value products (e.g. pharmaceutical uses).

2   Frameworks for identifying and managing food loss and waste in developed dairy chains

The Food Waste Hierarchy, proposed by Papargyropoulou et al. (2014) and described in detail in the following paragraphs, has been adopted by governments and industry as a framework for managing food waste. This is because it broadly outlines different management practices, with prevention of surplus food production as being the most desirable management practice, through to recovering soil nutrients and energy and, lastly, disposal (e.g. through landfill or wastewater).

There have been many iterations of the Food Waste Hierarchy (Jones et al., 2022; Redlingshöfer et al., 2020). The version proposed by Teigiserova et al. (2020) attempts to clarify some of the inconsistencies in the terminology and the order of preference (see Fig. 3). In Teigiserova et al.'s (2020) version of the hierarchy, the order of preference is as follows:

1. Prevention (remains the most desired management practice);
2. Reusing edible food waste in the human supply chain (e.g. through donations to food rescue organisations);
3. Reuse of inedible food wastes as animal feed;
4. Recycling of food waste streams to extract valuable materials (i.e. for non-food applications);
5. Recovering soil nutrients (e.g. composting which can be used again for agricultural production);

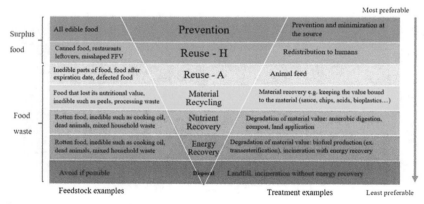

Figure 3 Food surplus and waste hierarchy. Source: Reproduced with permission from: Teigiserova et al. (2020).

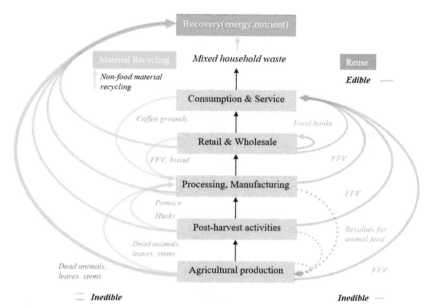

Figure 4 The circular economy framework for food surplus, waste and loss. Source: Reproduced with permission from: Teigiserova et al. (2020).

6 Energy recovery (e.g. production of biofuels); and
7 Disposal.

Teigiserova et al. (2020) apply the updated hierarchy to a typical food value chain, incorporating circular economy thinking (as shown in Fig. 4). The colours translate to possible management options for each stage of the chain. Green represents reuse in human (solid line) or animal feed (dotted line); yellow represents material recycling of non-food components; and orange represents recovery of energy or nutrients.

The framework by Teigiserova et al. (2020) provides a foundation that can be adapted to specific food sectors, including dairy. However, an issue with this framework is that recovered soil nutrients and energy are not directed back to agricultural production, meaning it is not truly 'circular', and it does not consider the recovery of water from FLW an important outcome. We propose the expanded framework in Fig. 5 to overcome these limitations and apply management practices available to developed dairy chains. In this framework, dairy products move from production to processing, retail or wholesale, consumption (e.g. in households, food service or institutions), recovery and then back to production. Additionally, there are multiple management practices that could be employed at each stage of the chain that would prevent the disposal of waste. These are summarised in Section 3.

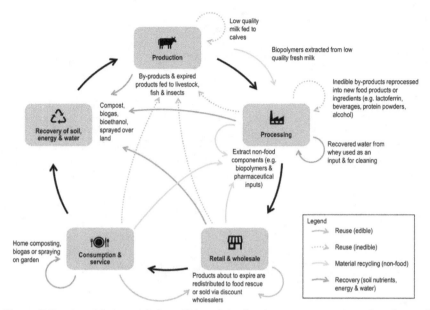

Figure 5 Proposed framework for utilising dairy food waste streams in a close-looped paradigm.

3 Key stages and management practices for reducing food loss and waste in dairy chains

3.1 Production

Production losses on farms in developed dairy chains are relatively low compared to regions with under-developed chains. The major source of on-farm loss is milk that does not meet minimum quality parameters due to issues such as bacterial contamination (Lore et al., 2005) or antibiotic residues (March et al., 2019). A common management practice is feeding this milk to calves or to other livestock species such as pigs.

According to the Food Waste Hierarchy, recycling of non-food components present in FLW is preferred when it cannot be reused as human or animal food. Biopolymers can be extracted from milk that does not meet minimum quality specifications, which can be used to manufacture bio-plastics (Jefferson et al., 2020). Although high production costs have been reported as a significant issue for bio-plastic production from dairy products (de Castro et al., 2022), there is some evidence that it can return a positive return on investment (Chalermthai et al., 2020). The negative environmental impacts of such bio-plastic production still remain because of the chemical by-products from the manufacturing process (Chalermthai et al., 2021; de Castro et al., 2022). This is discussed in the Section 3.2.

Other on-farm management practices for dealing with waste from milk production include recovering soil nutrients, energy and water from milk that does not meet the minimum specifications through anaerobic digestion (AD). These can be co-digested by mixing with dairy cattle effluent (i.e. manure) (Wu et al., 2011).

3.2 Processing

During the processing stage of the value chain, there can be significant waste due to the generation of by-products. Whey, which is the most significant by-product from the manufacturing of dairy products such as cheese and yoghurt, has historically been disposed of through wastewater treatment (Smithers, 2008). However, whey management options other than disposal include:

1 re-processing into new products;
2 as a feed source in livestock production;
3 recycling non-food components; and
4 recovering soil nutrients, energy and water.

Tsermoula et al. (2021) describe the emergence of more sustainable and profitable processing techniques to valorise whey. Fig. 6, which has been adapted from Juliano et al. (2017), illustrates different processing options and end products and how these align with the management practices outlined in Fig. 3:

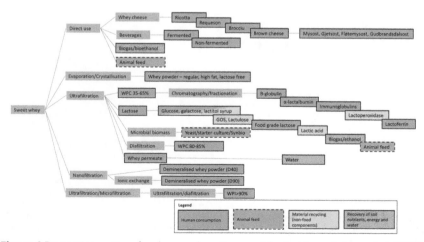

Figure 6 Processing options for cheese whey. Source: Adapted from: Juliano et al. (2017).

- Products that are intended for human consumption (green);
- Animal feed (green with a dotted outline);
- Material recycling of non-food components (yellow); and
- Recovery of soil nutrients, energy and water (orange).

While this is not an exhaustive list of processing techniques and end products, it highlights the many management options available to dairy processors.

The resulting products (e.g. powders or protein and lactose) can be used in a variety of industries such as nutraceuticals, infant formulas and confectionery products (Oliveira et al., 2020; Tsermoula et al., 2021). Lactoferrin is used as a nutritional supplement with anti-inflammatory properties and has been investigated as a potential treatment for severe coronavirus disease 2019 (COVID-19) (Chang et al., 2020). Lactose isolated from whey can be further fermented, or co-fermented with other inputs such as wheat, into ethanol. The ethanol can be used in alcoholic beverages, vinegars and biofuels (Ling, 2008; Murari et al., 2019; Parashar et al., 2016; Zotta et al., 2020). Some processing options, including forward osmosis/reverse osmosis and ultra-filtration/reverse osmosis, are cost-effective means of producing whey powder and recovering water (Aydiner et al., 2014; Meneses and Flores, 2016).

There are many processing options for whey, but as García-Flores et al. (2015) show, the optimal choice for processing whey depends on various socioeconomic and production factors, such as the quantity of whey being generated. Figure 7 illustrates the expected earnings before interest and tax (EBIT) from a range of processing options and the capacity ranges to make it financially viable. An additional consideration to make these financially viable is the capital and operating costs.

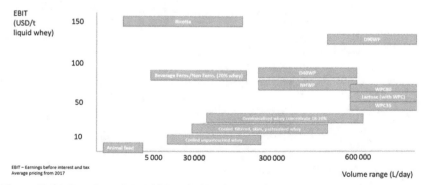

Figure 7 Options for value adding whey with consideration to the capacity range to make processing options viable and the expected earnings before interest and tax (EBIT). Source: Adapted from: Juliano (2021).

Other options that circular dairy systems can employ to minimise FLW during processing include reusing waste streams in animal industries. By-products, such as whey, are good sources of protein and energy and, to a lesser extent, micronutrients such as calcium and phosphorus (Schingoethe, 1976). This means they are an adequate feed source for ruminants, pigs, poultry, fish, horses, and rats (Sallam et al., 2022; Schingoethe, 1976). Ruminants and pigs can consume liquid and dry whey effectively, and dried whey has been shown to improve the quality of silage. Another option for diverting whey, and other dairy waste products, is by feeding it to *Hermetia illucens* larvae, commonly known as black soldier flies (BSF) (Hadj Saadoun et al., 2020). By feeding dairy FLW to BSF, the larvae can be fed to other livestock industries such as pigs, poultry and fish (Barragan-Fonseca et al., 2017). However, the feasibility of this will depend on the cost of transport and the price the processor can sell the whey to the animal industry.

Recycling non-food components from dairy FLW during processing includes extracting biopolymers and lactoperoxidase. Biopolymers extracted from whey, such as polylactic acid (PLA) and polyhydroxyalkanoates (PHA), can be used to manufacture bio-plastics, which can be a substitute for plastics derived from fossil fuels (Asunis et al., 2020). This can be achieved by processing liquid and powdered whey which has been shown to be economically feasible (Chalermthai et al., 2020). However, there are persistent environmental issues arising from processing, in particular, from the chemical input polyethylene glycol methyl ether methacrylate (PEGMA) (Chalermthai et al., 2021). Further research is needed to find alternative chemical processes that are environmentally sustainable. Lactoperoxidase has antimicrobial properties and has been recommended to reduce microbial levels in milk where pasteurisation or cold chain infrastructure is not available (Codex, 1991). It can also be combined with biopolymers to act as antimicrobial in edible packaging films (Zhang and Rhim, 2022). A study undertaken by Saravani et al. (2019) found that lactoperoxidase was an effective additive to biodegradable coatings for Gouda cheese products.

Recovery of soil nutrients, energy and water from dairy FLW is possible through composting and AD, e.g. for use in biogas or as compost (Hauser, 2017). Tait et al. (2021) estimate high energy recovery potential from dairy processing waste streams through AD (see Fig. 8), which can be coupled with effluent from dairy farms (Mostafa Imeni et al., 2019). This can contribute to offsetting the intensive energy requirements from processing and cooling milk (Mainardis et al., 2019). However, there are a number of factors related to the individual business that affect the viability of AD systems. This includes waste production volumes, space availability, infrastructure cost, energy costs, ability to utilise biogas energy, skilled labour, location and proximity to the processing plant (GHD, 2017). Utilising whey as a feedstock for bio-hydrogen has been explored

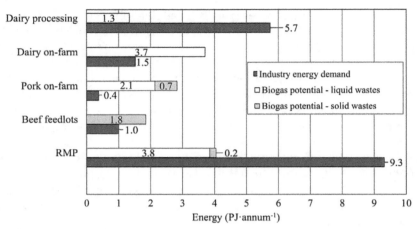

Figure 8 Energy requirements and biogas potential of animal industries in Australia. Note: RMP, red meat processing. Source: Reproduced with permission from: Tait et al. (2021).

as an alternative energy source (Patel et al., 2016). However, there are technical challenges to ensure the necessary biological parameters for anaerobic (co-) digestion are met and remain cost-effective (Tait et al., 2021). Application of whey to the land has been a common practice for whey. However, the correct dilution is required to prevent negative environmental and health outcomes (Ghaly et al., 2007). The soil type and landscape (e.g. the slope and proximity to rivers) can restrict the ability to spray whey onto the land (Watson et al., 1977).

3.3 Retail and wholesale

At the retail and wholesale stage of the chain, options for managing FLW include redistribution of edible food for human consumption, diverting food to animal feed, recycling non-food components and recovery of nutrients and energy.

Redistributing edible food is an important activity towards the goal of reducing food waste while tackling food insecurity (Reynolds et al., 2015). This includes redistributing soon-to-expire food via food rescue organisations such as food banks or discount wholesalers. For instance, in Australia, a company called 'Yume' (https://yumefood.com.au/) connects producers and processors with excess stock to retailers and food service businesses at a discounted rate. These options ensure food remains part of the human food chain and, in the case of food rescue, help address food insecurity. Further, with the consideration of tax incentives for businesses that donate excess food (KPMG, 2020), this will likely become a more viable option for businesses along the value chain.

Much like the options available to processors, feeding expired dairy products can also be diverted to animal feed (e.g. via BSF larvae production) (Khyade and Tamhane, 2021) or recovery through industrial composting facilities or AD. Extracting biopolymers (i.e. recycling) for use is technically feasible from expired milk products but is not a viable commercial solution currently (Jefferson et al., 2020).

3.4 Consumption and service

The consumption stage of the chain includes households, food service (e.g. restaurants) and institutions (e.g. hospitals and schools). Management options here are relatively similar to other stages of the value chain in that FLW can be redistributed to humans, used as animal feed, non-food components recycled and, lastly, soil nutrients, energy and water are recovered.

For food service and institutions, due to their relative scale, management options available are similar to retailers; i.e. redistribution of edible excess food to consumers; diversion to animal feed (e.g. BSF); and recovery through industrial composting facilities or AD.

For households, it is likely that significant reductions in food waste can be achieved, especially in developed countries like Australia (Ananda et al., 2021). Options include purchasing extended-shelf life or shelf-stable milk products, availability of smaller containers and improving planning behaviours in relation to purchasing and consumption so less is wasted (Williams et al., 2020). However, for waste that cannot be completely prevented and in-keeping with circular-economy-thinking, options include applying small or diluted amounts of milk to home gardens, composting and using small biogas digesters (Alexander et al., 2019). However, these options will probably not be available to all households as access to gardens, available space (e.g. for biogas digesters) and income may be limiting factors. Therefore, municipal waste management (solid waste) and wastewater treatment (for milk poured down the sink) would be needed to allow for the recovery of nutrients, energy and water.

3.5 Recovery of soil nutrients, energy and water

In this step, all of the nutrients not reused in human and animal feed, or processed to extract valuable materials, need to go through aerobic or anaerobic digestion to recover the remaining nutrients, energy and water. Again, in-keeping with circular economy thinking, these will be directed back as fertilisers for crops or as biofuel which, in the case of dairy production and processing, can be very energy intensive (Tait et al., 2021).

Management options like AD can be combined with effluent from dairy cattle. While not being classified as 'recovery' in the Food Waste Hierarchy, effluent from dairy cattle can also be used as feedstock for BSF (Myers et al.,

2014). This represents a significant opportunity for businesses that produce and process milk and also have access to other animal industries.

4 Improving decision-making in managing food loss and waste in dairy value chains

As highlighted, there are a number of management options for addressing dairy FLW. As per the Food Waste Hierarchy, a prescribed order of priority has been established for considering these management options. Where possible, the management options higher in the hierarchy should be exhausted before considering any in the lower tiers. For instance, the Food Waste Hierarchy suggests that AD should not be considered if feeding to pigs or other animals is an available option. This approach has been used by proponents of the Food Waste Hierarchy, as it provides guidance to decision-makers (e.g. governments, industry and consumers) about what would reduce the impacts of their waste, including from an environmental and social perspective (Papargyropoulou et al., 2014).

There is conflicting evidence with regard to which management options are the most effective, especially for management options in the middle tiers of the Food Waste Hierarchy (Redlingshöfer et al., 2020). Moult et al. (2018) found that, in order to minimise the GHG emissions from cheese products wasted at the UK retail level of the value chain, it was preferable to incinerate cheeses rather than to compost them. This conflicts with the generalised preferences outlined in the Food Waste Hierarchy. Many studies that examine the environmental impacts of FLW interventions do not consider multiple environmental outcomes (Redlingshöfer et al., 2020). This is important because interventions can have not only positive outcomes in some areas but also perverse outcomes in others. For example, Ridoutt et al. (2021) found that changes in diets to meet recommended dietary guidelines would reduce GHG emissions but would increase the water footprint. It is therefore important to consider multiple outcomes for each management option.

Businesses generating dairy FLW will need to identify the best management option factoring in the food product and quality, location of the business, the physical proximity of the next stage of the supply chain (e.g. piggeries, commercial composters, etc.), availability and access to technology (e.g. processing equipment), quantities of waste being generated and, most importantly, achieving minimum thresholds for a management practice that is financially viable (e.g. as shown in Fig. 7). Therefore, businesses need tools to support their FLW decision-making to identify the optimal choice for their operation, while considering the complexities and trade-offs between multiple financial and non-financial outcomes (Somlai, 2022). It is for this reason decision support tools (DSTs) have been described as a major priority for helping increase the sustainability of food systems (Béné et al., 2020).

4.1 Decision support tools used to manage dairy food loss and waste

DSTs are defined by Power (2002) as 'interactive computer-based systems that help people use computer communications, data, documents, knowledge, and models to solve problems and make decisions'. This definition covers a wide variety of technologies, including collaborative workspaces, financial models and record systems. DSTs have been referred to by many names including decision support systems, management information systems, executive information systems, business intelligence, collaborative systems and big data and cloud computing (Annosi et al., 2021; Power, 2002; Sprague, 1980).

Several DSTs have been used by decision-makers in the dairy sector to consider management options for their FLW. Eight DSTs relevant to dairy food products and waste management are summarised in Table 1. These DSTs utilise a variety of different analytical approaches including life cycle assessment (LCA)[1], life cycle costing (LCC)[2], material flow cost accounting (MCFA)[3] and benefit–cost analysis (BCA)[4].

There are a number of other DSTs that are publicly available to dairy and food business, including Simapro, Gabi, Umberto and OpenLCA (Silva et al., 2017). These all use analytical approaches such as LCA, LCC and MCFA to support decision-makers. Indeed, EPA's WARM utilises OpenLCA as the underlying programming to analyse the changes from FLW management. While these tools can be used to support more sustainable food systems, their primary aim is not to identify management options or reduce FLW issues; they are therefore not described in detail in this chapter. ERP tools like Systems Applications Products (SAP) and DEAR systems have inventory management capabilities, which can be used to identify how much food product stock needs to be managed. To make the best use of this information requires additional analysis, which can be customised or integrated with other systems to support FLW management. ERPs like SAP are not explored in detail but the authors recognise they can support businesses, especially through mass collection of inventory data. Lastly, some 'tools' are available to assist with quantifying and reporting FLW, e.g. WRAP's FLW Data Capture

[1] LCA is an evaluation of the 'inputs, outputs and the potential environmental impacts of a product system throughout its life cycle' (ISO 14040:2006).
[2] LCC is a 'methodology for systematic economic evaluation of life-cycle costs over a period of analysis, as defined in the agreed scope' (ISO 15686-5.2:2008).
[3] MCFA is a 'tool for quantifying the flows and stocks of materials in processes or production lines in both physical and monetary units' (ISO 14051:2011).
[4] BCA is a widely used economic technique to evaluate of the benefits and costs associated with an intervention, usually in monetary terms.

Table 1 Summary of decision support tools (DSTs) in dairy product and waste management

Name	Description	Reference	URL
Waste reduction model (WARM) tool (v15)	Uses life cycle assessment (LCA) and life cycle costing (LCC) to compare the environmental and economic impacts of managing waste material. This includes non-dairy and non-food materials. It is available in an Excel spreadsheet and an open-source openLCA program.	(EPA, 2019)	https://www.epa.gov/warm/versions-waste-reduction-model-warm#15 (accessed 30 March 2022)
Food loss and waste value calculator (Beta v1.1)	Uses LCA to compare impacts on environmental outcomes (GHG emissions). The Excel-based calculator is designed to complement the Food Loss Waste Protocol to assist decision-makers across the food chain to consider different management options.	(WRI, 2019)	https://flwprotocol.org/why-measure/food-loss-and-waste-value-calculator/ (accessed 30 March 2022)
Insights Engine	An interactive platform to access data and results about food loss and waste in the USA. The web-based components in the Insights Engine include a. Waste monitor – database of FLW in the USA. b. Solutions Database and Directory – quantified benefits of different management options. c. Impact Calculator – an interactive tool to compare environmental and social outcomes in the changes in FLW end destinations. d. Policy Tracker – an interactive map with updated information about relevant state and federal FLW policies.There is a combination of LCA and BCA in the Insights Engine.	(Powell and Curtis, 2020a, b)	https://insights.refed.org/ (accessed 30 March 2022)
Whey distribution chain optimisation model	Allows users to optimise the selection of whey processing equipment and allocation between different cheese manufacturers in a region and candidate whey collection and concentration centres. The model selects the most adequate location for the drying facilities (if total concentration volumes are sufficient). This is a linear programming optimisation model, which uses an Excel-based platform and generates a Google Earth output. The model also provides the return of investment to the region for capital equipment involved. It includes options for using secondhand equipment or drying facilities with spare capacities.	(García-Flores et al., 2015)	NA

Tool	Description	Reference	URL
Benders decomposition for a reverse logistics network	Builds on the whey distribution chain optimisation model by Garcia-Flores et al. (2015) by accounting for whey supply uncertainty. The former model relies on two sequential stages of (1) identifying facility locations and (2) allocating whey between sites depending on the amount available day-to-day. This approach does not account for uncertainty in whey production. The model using classical HFLP with stochastic demands Benders Decomposition accounts for this uncertainty and indicates a reduction in costs by 28%.	(Esmaeilbeigi et al., 2021)	NA
Whey utilisation tool (whey tool)	The Whey Tool was developed for the Australian dairy processing sector and uses BCA to estimate the capital and operating expenditure required to process whey into new products. The web-based DST allows users to evaluate the financial viability of selling a range of whey-derived products.	(Hauser, 2013)	https://wheytool.dairyaustralia.com.au/ Demonstration (accessed 30 March 2022)
FOod side flow Recovery LIfe cycle Tool (FORKLIFT)	FORKLIFT is an Excel-based DST that helps businesses think through the costs and emissions associated with FLW streams from their operation. It is tailored to priority food waste streams in Europe, including whey permeate. Others include apple pomace, brewers spent grain, rape seed press cake, tomatoes pomace and abattoir blood.	(Davis et al., 2019)	https://eu-refresh.org/forklift.html (accessed 30 March 2022)
GISWASTE	GISWASTE uses AHP, as well as LCA, BCA and GIS, to evaluate the technical requirements, and economic and environmental impacts of different waste management options for dairy, meat and vegetable waste producers. Users of this tool are waste management organisations looking to source from local producers/processors. GISWASTE uses Microsoft Sharepoint, Access and ArcGIS.	(San Martin et al., 2017)	NA
Dynamic Industry Resource Efficiency Calculation Tool (DIRECT)	DIRECT was co-developed, via a researcher partnership with food manufacturers, and later commercialised (as a Fight Food Waste CRC project). It uses mass balance to track and visualise a food organisation's product and non-product material flows. As well as using material flow cost accounting (MFCA) principles to support businesses to calculate the 'true' cost of waste, by accounting for the input and operational costs attributable to mass that is lost in a process or along the supply chain. It is a cloud-based DST and aligns with the FLW Standard (developed by the FLW Protocol) as well as the MFCA ISO 14051 (and its derivatives 14052 and 14053). The user defines the scope of the assessment and can compare multiple scenarios.	(Empauer, 2022; Verghese et al., 2018)	https://empauer.com/solutions/direct/ (accessed 30 March 2022)

(Continued)

Table 1 (*Continued*)

Name	Description	Reference	URL
Ecodex	Ecodex is a cloud-based lifecycle product management DST co-developed by Nestle and Selerant. The tool utilises LCA databases to estimate the environmental footprint of products in a range of sectors including packaging. For food manufacturers, it can account for food losses (e.g. spillage and the generation of by-/co-products) and the relative contribution of these to environmental outcomes such as GHG emissions and water consumption. Ecodex helps product managers and businesses during the product design stage and, therefore, can be used to help prevent or minimise FLW.	(Schenker et al., 2014)	https://empauer.com/solutions/ecodex/ (accessed 17 June 2022)
Integral Dairy Valorisation Model (IDVM)	IDVM uses a linear programming model to identify the optimum allocation of raw dairy materials available to a dairy processing company. The tool uses a specialised data analysis program and was designed in partnership with Friesland Campina in the Netherlands.	(Banaszewska et al., 2014; Banaszewska et al., 2013)	NA

LCA, life cycle assessment; LCC, life cycle cost; GHG, greenhouse gas; BCA, benefit–cost analysis; HFLP, hierarchical facility location problems; AHP, analytical hierarchy process; GIS, geographic information system; MFCA, material flow cost accounting; NA, not applicable.

Sheet (WRAP, 2022). These tools were not considered in this chapter, though we acknowledge the importance of standardised reporting frameworks to assist with managing FLW.

The following paragraphs provide an overview of each of these DSTs. It takes into consideration the type/s of dairy products and product categories; location specificity (e.g. state, country or regional data); sustainability decision criteria (i.e. social, economic and environmental); stage/s of the value chain; and FLW management options. These characteristics are outlined in Table 2. As highlighted in Table 2, the eight DSTs provide different levels of support and specificity to dairy products and waste streams. Table 2 also illustrates that no single DST adequately covers all stages of the value chain or all management options.

For instance, on the one hand, ReFED's Insights Engine provides guidance on economic, social and environmental outcomes in relation to many food products, from primary production through to consumption. It covers management practices at all stages of the Food Waste Hierarchy (i.e. prevention through to disposal). However, users are not able to separate information specific to the dairy sector, either as a category or as individual products. Users must rely on the broader sub-category of 'dairy and eggs'. Therefore, this DST will provide general guidance across a number of areas, but it is not able to provide information specific to dairy products.

On the other hand, the Integral Dairy Valorisation Model (IDVM) developed by Banaszewska et al. (2014) provides highly targeted information about 54 individual dairy by-/co-products and optimises raw materials based on milk supply, market demand, production and transportation. The model – which is specifically targeted to milk processing companies – optimises profit based on the further processing of waste streams into new value-added products. Therefore, it does not consider other FLW management options such as feeding to livestock, anaerobic digestions and so on. Also, IDVM only accounts for the economic dimension of sustainability, compared to other DSTs which also incorporate social and environmental outcomes in their analysis.

Based on the information outlined in Tables 1 and 2, the options to manage and valorise dairy FLW are captured to different degrees in the currently available DSTs, and relevance to different stages of value chain is summarised in Table 3. Further research is needed to develop tools that support dairy value chain actors to valorise dairy FLW.

4.2 Factors that need to be considered

It is becoming increasingly necessary for food businesses at any stage of the value chain, including in the dairy sector, to move towards close-looped chains. Therefore, dairy businesses need to consider the full suite of options available

Table 2 Characteristics of existing DSTs

Name	Dairy product specificity	Location specificity	Sustainability decision criteria	Stage of the chain	Management options considered
WARM tool	• 'Dairy' proxy • Mixed food waste	• USA (national and state data)	**Economic:** • Labour hours • Wages • Taxes **Environmental:** • GHG emissions (CO_2-e) • Energy	• Waste management	• Prevention • Recycling • Anaerobic digestion • Composting • Combustion • Landfill
Food loss and waste value calculator	• Cheese • Milk • Yoghurt • 'Dairy' proxy (average of cheese, milk and dairy)	• Global and regional data	**Environmental:** • Climate change (CO_2-e) • Water scarcity footprint (m^3-e) • Soil quality index (points) • Eutrophication (phosphorous and nitrogen) **Social:** • Energy (kcal) • Protein (g) • Carbohydrates (g) • Fibre (g) • Micronutrients (mg)	• Production • Processing • Distribution • Retail • Households • Waste management	• Animal feed • Bio-based materials/biochemical processing • Co-digestion/anaerobic digestion • Composting/aerobic processes • Controlled combustion • Land application • Not harvested/ploughed in • Landfill • Refuse/discards/litter • Sewer/wastewater treatment

Tool	Product	Geography/data	Metrics	Value chain stage	End-of-life options
Insights Engine	• 'Dairy and eggs' proxy	• USA (national and state data)	**Economic:** • Cost & benefit **Environmental:** • Climate change (CO$_2$-e) • Water footprint (m^{3}e) **Social:** • Meals wasted/saved (equivalents) • Jobs	• Production • Processing • Distribution • Retail • Food service • Households	• Prevention • Donation • Industrial uses • Animal feed • Anaerobically digested • Composted • Not harvested • Incineration • Land application • Landfill • Dumping • Sewer
Whey distribution chain optimisation model	• Whey	• Non-specific (case study conducted in Brazil (published) as well as Argentina, Colombia and Uruguay (unpublished))	**Economic:** • Cost • Return on investment **Environmental:** • Regional map for whey waste avoidance	• Processing	• Further processing into powders • Option for further processing into dairy consumer products such as beverages and dairy snacks
Benders decomposition for a reverse logistics network	• Whey	• Non-specific [case study conducted in Brazil (published)]	**Economic:** • Cost	• Processing	• Further processing into powders

(Continued)

Table 2 (*Continued*)

Name	Dairy product specificity	Location specificity	Sustainability decision criteria	Stage of the chain	Management options considered
Whey tool	Whey-derived products, specifically: • Liquid raw whey • Liquid pasteurised Whey • Demineralised 40% whey (liquid and powder) • WPC 35 (liquid and powder) • Whey Powder • Ricotta Cheese • Whey Cream • Whey Permeate • Ricotta Whey	• South-east Australia • New Zealand • Europe/North America	**Economic:** • Years-to-payback	• Processing	• Further processing into food products
FORKLIFT	• Whey permeate	• EU (regional and national data)	**Economic:** • Cost **Environmental:** • GHG emissions $(CO_2\text{-e})$	• Processing	• Further processing into whey permeate powder • Animal feed • Anaerobic digestion • Bioethanol production

GISWASTE	• Dairy	• Basque country, Spain	**Economic:** • Cost • Revenue/income **Environmental:** • Distance travelled • GHG emissions (CO_2-e) • Water footprint (m^3-e) • Wastewater generated • Eutrophication potential (kg P-e) • Biogas/animal feed produced	• Waste management	• Animal feed • Anaerobic digestion
DIRECT	• >50 dairy-related product groups, classes and subclasses outlined in the UN Central Product Classification (CPC), Version 2.1	• Non-specific	**Economic:** • 'True' cost of waste **Environmental:** • Tonnes of FLW • GHG emissions (CO_2-e) • Non-renewal energy and minerals • Water footprint (m^3-e) • Land use • Ecosystem quality	• Processing • Distribution • Retail • Food service	• Prevention • Donation • Industrial uses • Animal feed • Anaerobically digested • Composted • Not harvested • Incineration • Land application • Landfill • Dumping

(Continued)

Table 2 (*Continued*)

Name	Dairy product specificity	Location specificity	Sustainability decision criteria	Stage of the chain	Management options considered
Ecodex	• 33 dairy products based on CPC product types are available in the life cycle inventory database	• Geographic specificity can be at the global, regional, country and provincial level	**Environmental:** • Tonnes of FLW • GHG emissions (CO_2-e) • Non-renewal energy and minerals • Water footprint (m3-e) • Land use • Ecosystem quality	• Processing (Ecodex is primarily used for decision-makers at the processing stage but represents impacts across the product lifecycle; from production to retail)	• Prevention
IDVM	• 54 individual dairy (by-/co-) products	• The Netherlands (FrielandCampina)	**Economic:** • Profit	• Processing	• Prevention • Further processing of by-products into powders and individual food components

Table 3 FLW DST that are available to decision-makers at different stages of the dairy value chain

	Primary production	Processing	Retail and wholesale	Consumption and service	Recovery/ waste management
WARM Tool					•
Food loss and waste value calculator	•	•	•	•	•
Insights Engine	•	•	•	•	
Whey distribution chain optimisation model		•			
Benders decomposition for a reverse logistics network		•			
Whey tool		•			
FORKLIFT		•			
GISWASTE					•
DIRECT	•	•	•	•	
Ecodex		•			
IDVM		•			

(i.e. reduce, reuse, recycle and recover) to manage waste streams. However, new DSTs need to account for the specific problems facing the dairy sector and consider their stage of the value chain, location and proximity to other industries, and the cost and access to technology. A list of example research questions that would need to be addressed during the decision-making process for managing whey, and therefore need to be integrated into a DST, is as follows:

- How much whey is currently produced, including seasonal variations?
- What are the current practices in place to utilise whey and the associated costs, revenue, environmental and social impacts?
- Is there a demand for alternative products (e.g. food products, animal feed, biogas and feedstock)?
- What management pathways are being considered (e.g. processing in-house with existing infrastructure, investing in new infrastructure, outsourcing, selling raw materials to other businesses or combining with other businesses' waste products)?

- What are the minimum volumes/capacity required for each management option being considered?
- What are the associated transportation requirements?
- What investment would be needed to self-process and what would be the return on investment?
- What are the benefits and costs to transform/utilise the whey?
- What are the environmental benefits and trade-offs for each management option (e.g. GHG emissions, water footprint, etc.)?
- What are the social benefits and trade-offs for each management option (e.g. meals saved, nutrients, etc.)?
- What type of whey, and what is the quality and composition of the whey?

5 Conclusion

FLW in dairy chains is a significant issue for society but there are many opportunities to re-purpose and valorise this resource by considering circular economy principles. These include feeding more humans, using the waste as animal feed (including livestock, fish and insects), recycling non-food components, or recovering soil nutrients, energy and water. This chapter outlines several management practices that are available to decision-makers in the dairy sector. The proposed framework for utilising dairy food waste streams in a close-looped paradigm (Fig. 5) is an important contribution to the broader discussion about transitioning to circular food systems. However, knowing which practice or technology may be best suited to individual businesses and economically, environmentally and socially sustainable can be a challenge for the decision-makers involved.

DSTs can help business decision-makers to navigate the Food Waste Hierarchy and complex broader issues related to FLW. Current DSTs address some of the needs of decision-makers but do not provide full coverage of decision support for dairy businesses. Therefore, further work is required to develop DSTs that integrate various aspects of the existing tools into one tool. The ideal DST would enable companies or company groups (e.g. cooperatives or joint ventures).

Additionally, as discussed at the beginning of this chapter, the challenges facing different regions vary. This chapter highlights the unique issues and opportunities facing regions with developed dairy chains. Further research is needed to better understand the unique challenges and potential solutions facing the regions with underdeveloped dairy chains.

By providing the tools and information to decision-makers through the dairy value chain, adopting practices that align with the circular economy will be more achievable and contribute to the goal of halving global FLW.

6 Acknowledgements

This work has been supported by the following organisations: (1) Fight Food Waste Cooperative Research Centre whose activities are funded by the Australian Government's Cooperative Research Centre Program; (2) the Commonwealth Scientific and Industry Research Organisation (CSIRO); and (3) the University of Adelaide. Thanks to Paul-Antoinne Bontinck (Lifecycles) for providing access to the National Food Waste Baseline data and to Victor Barichello (Empauer) and Dejan Popovic (Selerant) and Alister Hall and Simon Lockrey (RMIT) for their feedback to this chapter.

7 Where to look for further information

There has been a considerable amount of information published on the topic of whey value addition, including:

- Smithers, G. W. (2008). Whey and whey proteins—From 'gutter-to-gold', *International Dairy Journal* 18(7), 695–704. https://doi.org/10.1016/j .idairyj.2008.03.008.
- Oliveira, D., Fox, P. and O'Mahony, J. A. (2020). Chapter 4: Byproducts from Dairy Processing, In: Simpson, B. K., Aryee, A. N. A. and Toldrá, F. (Eds.), *Byproducts from Agriculture and Fisheries*. John Wiley & Sons Ltd, West Sussex, UK. https://doi.org/10.1002/9781119383956.ch4.
- Fenwick, R., Harper, J., Hobman, P., Huffman, L., Kirkpatrick, K., MacGibbon, J., Marshall, K., Matthews, M., Wilson, A. and Woodhams, D. (2014). Whey to go: whey protein concentrate: a New Zealand success story. Ngaio Press, Martinborough, New Zealand, https://nzifst.org.nz/page-18143#:~ :text=Whey%20to%20Go%20is%20the,%2C%20Robin%20Fenwick %2C%20Arthur%20Wilson.&text=A%20fascinating%20story%20about %20how%20industrial%20innovation%20really%20works%20in %20practice.
- Tetra Pak (2015). Chapter 15 - Whey Processing, In: Bylund, G. (Ed.) Dairy Processing Handbook. Tetra Pak, Pully, Switzerland.

Australia's peak body for the dairy industry recently published a food waste sector action plan, which is available here: https://www.dairyaustralia.com.au /manufacturing-support/manufacturing-sustainability/dairy-sector-food-waste -action-plan#:~:text=The%20Dairy%20Food%20Waste%20Action,across %20the%20dairy%20supply%20chain.

For further information about decision support tools refer to Table 1.

8 References

Alexander, S., Harris, P. and McCabe, B. K. (2019). Biogas in the suburbs: an untapped source of clean energy?, *Journal of Cleaner Production* 215, 1025–1035. https://doi.org/10.1016/j.jclepro.2019.01.118.

Ananda, J., Karunasena, G. G. and Pearson, D. (2021). *Priority Behaviours for Interventions to Reduce Household Food Waste in Australia*. Fight Food Waste Cooperative Research Centre (FFWCRC), Adelaide, Australia.

Annosi, M. C., Brunetta, F., Bimbo, F. and Kostoula, M. (2021). Digitalization within food supply chains to prevent food waste. Drivers, barriers and collaboration practices, *Industrial Marketing Management* 93, 208–220. https://doi.org/10.1016/j.indmarman.2021.01.005.

Asunis, F., De Gioannis, G., Dessì, P., Isipato, M., Lens, P. N. L., Muntoni, A., Polettini, A., Pomi, R., Rossi, A. and Spiga, D. (2020). The dairy biorefinery: integrating treatment processes for cheese whey valorisation, *Journal of Environmental Management* 276, 111240. https://doi.org/10.1016/j.jenvman.2020.111240.

Aydiner, C., Sen, U., Topcu, S., Sesli, D., Ekinci, D., Altınay, A. D., Ozbey, B., Koseoglu-Imer, D. Y. and Keskinler, B. (2014). Techno-economic investigation of water recovery and whey powder production from whey using UF/RO and FO/RO integrated membrane systems, *Desalination and Water Treatment* 52(1–3), 123–133. https://doi.org/10.1080/19443994.2013.786655.

Banaszewska, A., Cruijssen, F., Claassen, G. D. H. and van der Vorst, J. G. A. J. (2014). Effect and key factors of byproducts valorization: the case of dairy industry, *Journal of Dairy Science* 97(4), 1893–1908. https://doi.org/10.3168/jds.2013-7283.

Banaszewska, A., Cruijssen, F., van der Vorst, J. G. A. J., Claassen, G. D. H. and Kampman, J. L. (2013). A comprehensive dairy valorization model, *Journal of Dairy Science* 96(2), 761–779. https://doi.org/10.3168/jds.2012-5641.

Barragan-Fonseca, K. B., Dicke, M. and van Loon, J. J. A. (2017). Nutritional value of the black soldier fly (Hermetia illucens L.) and its suitability as animal feed-a review, *Journal of Insects as Food and Feed* 3(2), 105–120. https://doi.org/10.3920/JIFF2016.0055.

Béné, C., Fanzo, J., Haddad, L., Hawkes, C., Caron, P., Vermeulen, S., Herrero, M. and Oosterveer, P. (2020). Five priorities to operationalize the EAT-Lancet Commission report, *Nature Food* 1(8), 457–459. https://doi.org/10.1038/s43016-020-0136-4.

Chalermthai, B., Ashraf, M. T., Bastidas-Oyanedel, J. R., Olsen, B. D., Schmidt, J. E. and Taher, H. (2020). Techno-economic assessment of whey protein-based plastic production from a co-polymerization process, *Polymers (Basel)* 12(4), 847. https://doi.org/10.3390/polym12040847.

Chalermthai, B., Giwa, A., Schmidt, J. E. and Taher, H. (2021). Life cycle assessment of bioplastic production from whey protein obtained from dairy residues, *Bioresource Technology Reports* 15, 100695. https://doi.org/10.1016/j.biteb.2021.100695.

Chang, R., Ng, T. B. and Sun, W. Z. (2020). Lactoferrin as potential preventative and adjunct treatment for COVID-19, *International Journal of Antimicrobial Agents* 56(3), 106118. https://doi.org/10.1016/j.ijantimicag.2020.106118.

Chen, C., Chaudhary, A. and Mathys, A. (2020). Nutritional and environmental losses embedded in global food waste, *Resources, Conservation and Recycling* 160, 104912. https://doi.org/10.1016/j.resconrec.2020.104912.

Codex (1991). Guidelines for the preservation of raw milk by use of the lactoperoxidase system. Available at: http://siweb1.dss.go.th/standard/Fulltext/codex/CXG_013E .pdf. CAC/GL 13-1991. Codex Alimentarius Commission, Rome, Italy.

Davis, J., Holtz, E., Metcalf, P., De Menna, F., Scherhaufer, S., Gollnow, S., Colin, F., García Herrero, L., Vittuari, M. and Östergren, K. (2019). Valorisation models - Food side flow recovery lifecycle tool (FORKLIFT), REFRESH. Available at: https://eu-refresh .org/forklift.html.

de Castro, T. R., de Macedo, D. C., de Genaro Chiroli, D. M., da Silva, R. C. and Tebcherani, S. M. (2022). The potential of cleaner fermentation processes for bioplastic production: A narrative review of polyhydroxyalkanoates (PHA) and polylactic acid (PLA), Journal of Polymers and the Environment 30(3), 810–832. https://doi.org/10 .1007/s10924-021-02241-z.

de Wit, J. N. (2001). Lecturer's handbook on whey and whey products. Available at: http:// ewpa.euromilk.org/fileadmin/user_upload/Public_Documents/EWPA_Publications/ Lecturer_s_Handbook_on_Whey.pdf. European Whey Products Association (EWPA), Brussels, Belgium.

Empauer (2022). DIRECT. Available at: https://empauer.com/solutions/direct/ (Accessed 16 March 2022).

EPA (2019). Waste reduction model (WARM) tool. Available at: https://www.epa.gov/ warm/versions-waste-reduction-model-warm#15. Environmental Protection Agency (EPA), Washington DC.

Esmaeilbeigi, R., Middleton, R., García-Flores, R. and Heydar, M. (2021). Benders decomposition for a reverse logistics network design problem in the dairy industry, Annals of Operations Research. https://doi.org/10.1007/s10479-021-04309-4.

FAO (2013). Food wastage footprint. Impact on natural resources. Available at: http:// www.fao.org/publications/card/en/c/000d4a32-7304-5785-a2f1-f64c6de8e7a2/. Food and Agriculture Organization, Rome, Italy.

FIAL (2021). The National Food Waste Strategy Feasibility Study–Final Report. Available at: https://www.fial.com.au/sharing-knowledge/food-waste. Food Innovation Australia Ltd (FIAL), Werribee, Australia.

Gac, A., Béline, F., Bioteau, T. and Maguet, K. (2007). A French inventory of gaseous emissions (CH4, N2O, NH3) from livestock manure management using a mass-flow approach, Livestock Science 112(3), 252–260. https://doi.org/10.1016/j.livsci.2007 .09.006.

García-Flores, R., Martins, R., de Souza Filho, O. V., González, M., Mattos, C., Rosenthal, A. and Juliano, P. (2015). A novel facility and equipment selection model for whey utilisation: A Brazilian case study, Computers and Electronics in Agriculture 117, 127–140. https://doi.org/10.1016/j.compag.2015.07.016.

Ghaly, A. E., Mahmoud, N. S., Rushton, D. G. and Arab, F. (2007). Potential environmental and health impacts of high land application of cheese whey, American Journal of Agricultural and Biological Sciences 2(2), 106–117. https://doi.org/10.3844/ajabssp .2007.106.117.

GHD (2017). Anaerobic digestion as a treatment and energy recovery technology for dairy processing waste streams. Available at: https://www.dairyaustralia.com.au/ resource-repository/2020/12/03/anaerobic-digestion-as-a-treatment-and-energy -recovery-technology-for-dairy-processing-waste-streams#.YlkmzuhByUk. Dairy Australia, Melbourne, Australia.

Gustavsson, J., Cederberg, C., Sonesson, U. and Meybeck, A. (2011). Global food losses and food waste: Extent, causes and prevention. Available at: https://www.fao.org/3/i2697e/i2697e.pdf. Food and Agriculture Organization, Rome, Italy.

Hadj Saadoun, J., Montevecchi, G., Zanasi, L., Bortolini, S., Macavei, L. I., Masino, F., Maistrello, L. and Antonelli, A. (2020). Lipid profile and growth of black soldier flies (Hermetia illucens, Stratiomyidae) reared on by-products from different food chains (Hermetia illucens, Stratiomyidae), *Journal of the Science of Food and Agriculture* 100(9), 3648-3657. https://doi.org/10.1002/jsfa.10397.

Hanson, C., Lipinski, B., Robertson, K., Dias, D., Gavilan, I., Gréverath, P., Ritter, S., Fonseca, J., VanOtterdijk, R. and Timmermans, T. (2016). *Food Loss and Waste Accounting and Reporting Standard*.

Hanson, C. and Mitchell, P. (2017). The business case for reducing food loss and waste, *Champions of 12.3*. Available at: https://champions123.org/sites/default/files/2020-08/business-case-for-reducing-food-loss-and-waste.pdf. Washington, DC.

Hauser, J. (2013). Dairy Australia whey model documentation. Available at: https://wheytool.dairyaustralia.com.au/Content/pdf/Whey%20model%20Documentation%20July%202018%20v1.pdf. Dairy Australia, Melbourne, Australia.

Hauser, J. (2017). *Analysis of South Australia Whey Processing Options*. Xcheque, Australia.

Jefferson, M. T., Rutter, C., Fraine, K., Borges, G. V. B., de Souza Santos, G. M., Schoene, F. A. P. and Hurst, G. A. (2020). Valorization of sour milk to form Bioplastics: Friend or foe?, *Journal of Chemical Education* 97(4), 1073-1076. https://doi.org/10.1021/acs.jchemed.9b00754.

Jones, R. E., Speight, R. E., Blinco, J. L. and O'Hara, I. M. (2022). Biorefining within food loss and waste frameworks: A review, *Renewable and Sustainable Energy Reviews* 154, 111781. https://doi.org/10.1016/j.rser.2021.111781.

Juliano, P. (2021). Reinforcing South American communities through whey value addition, IDF Global Dairy Conference 13-15 October 2021. Copenhagen, Denmark.

Juliano, P., González, M. A., Castells, M. L., Di Risio, J., García-Flores, R., Rosenthal, A., Silva, C. M., Krolow, A. C. R., Zoccal, R., Walter, E., Dutra, A., Caicedo, I. B., Castañeda, C. G. G., Peñaranda, R. Q., Trujillo, R. S., Valencia, J. U. S., Álvarez, L. F. G., López, T., Jorcin, S., Miraballes, M., Mattos, C., Avellaneda, Y. A. and Salazar, P. V. (2017). *Whey Value Addition* (1st edn.). Instituto Nacional de Tecnología Industrial, San Martin, Argentina - INTI.

Khyade, V. B. and Tamhane, A. (2021). Utilization of the methanol maceratives of pre-pupal stages of the black soldier fly, Hermetia illucens L. (Diptera: Stratiomyidae) for inhibition of bacterial growth', *Uttar Pradesh Journal of Zoology*, 109-118.

Kirchherr, J., Reike, D. and Hekkert, M. (2017). Conceptualizing the circular economy: An analysis of 114 definitions, *Resources, Conservation and Recycling* 127, 221-232. https://doi.org/10.1016/j.resconrec.2017.09.005.

Knapp, J. R., Laur, G. L., Vadas, P. A., Weiss, W. P. and Tricarico, J. M. (2014). Invited review: Enteric methane in dairy cattle production: Quantifying the opportunities and impact of reducing emissions, *Journal of Dairy Science* 97(6), 3231-3261. https://doi.org/10.3168/jds.2013-7234.

KPMG (2020). A national food waste tax incentive. Available at: https://assets.kpmg/content/dam/kpmg/au/pdf/2020/national-food-waste-tax-incentive.pdf. KPMG, Sydney, Australia.

Kummu, M., de Moel, H., Porkka, M., Siebert, S., Varis, O. and Ward, P. J. (2012). Lost food, wasted resources: Global food supply chain losses and their impacts on freshwater, cropland, and fertiliser use, *Science of the Total Environment* 438, 477–489. https://doi.org/10.1016/j.scitotenv.2012.08.092.

Lade, S. J., Steffen, W., de Vries, W., Carpenter, S. R., Donges, J. F., Gerten, D., Hoff, H., Newbold, T., Richardson, K. and Rockström, J. (2020). Human impacts on planetary boundaries amplified by Earth system interactions, *Nature Sustainability* 3(2), 119–128. https://doi.org/10.1038/s41893-019-0454-4.

Ling, K. C. (2008). *Whey to Ethanol: A Biofuel Role for Dairy Cooperatives?* U.S. Department of Agriculture (USDA), Washington D.C.

Lore, T. A., Omore, A. O. and Staal, S. J. (2005). Types, Levels and Causes of Post-harvest Milk and Dairy Losses in Sub-Saharan Africa and the Near East: Phase Two Synthesis Report.

Mainardis, M., Flaibani, S., Trigatti, M. and Goi, D. (2019). Techno-economic feasibility of anaerobic digestion of cheese whey in small Italian dairies and effect of ultrasound pre-treatment on methane yield, *Journal of Environmental Management* 246, 557–563. https://doi.org/10.1016/j.jenvman.2019.06.014.

March, M. D., Toma, L., Thompson, B. and Haskell, M. J. (2019). Food waste in primary production: Milk loss with mitigation potentials, *Frontiers in Nutrition* 6, 173. https://doi.org/10.3389/fnut.2019.00173.

Meneses, Y. E. and Flores, R. A. (2016). Feasibility, safety, and economic implications of whey-recovered water in cleaning-in-place systems: A case study on water conservation for the dairy industry, *Journal of Dairy Science* 99(5), 3396–3407. https://doi.org/10.3168/jds.2015-10306.

Mostafa Imeni, S., Pelaz, L., Corchado-Lopo, C., Maria Busquets, A., Ponsá, S. and Colón, J. (2019). Techno-economic assessment of anaerobic co-digestion of livestock manure and cheese whey (Cow, Goat & Sheep) at small to medium dairy farms, *Bioresource Technology* 291, 121872. https://doi.org/10.1016/j.biortech.2019.121872.

Moult, J. A., Allan, S. R., Hewitt, C. N. and Berners-Lee, M. (2018). Greenhouse gas emissions of food waste disposal options for UK retailers, *Food Policy* 77, 50–58. https://doi.org/10.1016/j.foodpol.2018.04.003.

Murari, C. S., da Silva, D. C. M. N., Schuina, G. L., Mosinahti, E. F. and Del Bianchi, V. L. (2019). Bioethanol production from dairy industrial coproducts, *Bioenergy Research* 12(1), 112–122. https://doi.org/10.1007/s12155-018-9949-5.

Myers, H. M., Tomberlin, J. K., Lambert, B. D. and Kattes, D. (2014). Development of black soldier fly (Diptera: Stratiomyidae) larvae fed dairy manure, *Environmental Entomology* 37(1), 11–15. https://doi.org/10.1093/ee/37.1.11.

Oliveira, D., Fox, P. and O'Mahony, J. A. (2020). Chapter 4: Byproducts from dairy processing. In: Simpson, B. K., Aryee, A. N. A. and Toldrá, F. (Eds). *Byproducts from Agriculture and Fisheries*. John Wiley & Sons Ltd, West Sussex, UK.

Papargyropoulou, E., Lozano, R., Steinberger, K., Wright, N. and Ujang, Zb. (2014). The food waste hierarchy as a framework for the management of food surplus and food waste, *Journal of Cleaner Production* 76, 106–115. https://doi.org/10.1016/j.jclepro.2014.04.020.

Parashar, A., Jin, Y., Mason, B., Chae, M. and Bressler, D. C. (2016). Incorporation of whey permeate, a dairy effluent, in ethanol fermentation to provide a zero waste solution

for the dairy industry, *Journal of Dairy Science* 99(3), 1859–1867. https://doi.org
/10.3168/jds.2015-10059.

Patel, A. K., Vaisnav, N., Mathur, A., Gupta, R. and Tuli, D. K. (2016). Whey waste as potential feedstock for biohydrogen production, *Renewable Energy* 98, 221–225. https://doi
.org/10.1016/j.renene.2016.02.039.

Powell, C. and Curtis, P. (2020a). Insights engine food waste monitor - 2020 methodology. Available at: https://insights.refed.org/uploads/documents/refed_insights_engine
food_waste_monitormethodology_vfinal_2021.04.12.pdf?_cchid=4f1d745d5f5
0363e83869ef5f3b2e962. ReFED, New York.

Powell, C. and Curtis, P. (2020b). Insights engine solutions database - 2020 methodology. Available at: https://insights.refed.org/uploads/documents/refed-insights-engine
-solution-database-methodology-vfinal2021-05-27.pdf?_cchid=ccf71f4eacfac58
1ad228da51c320fd1. ReFED, New York.

Power, D. J. (2002). *Decision Support Systems: Concepts and Resources for Managers.* Greenwood Publishing Group, CT.

Redlingshöfer, B., Barles, S. and Weisz, H. (2020). Are waste hierarchies effective in reducing environmental impacts from food waste? A systematic review for OECD countries, *Resources, Conservation and Recycling* 156, 104723. https://doi.org/10
.1016/j.resconrec.2020.104723.

Reynolds, C. J., Piantadosi, J. and Boland, J. (2015). Rescuing food from the organics waste stream to feed the food insecure: An economic and environmental assessment of Australian food rescue operations using environmentally extended waste input-output analysis, *Sustainability* 7(4), 4707–4726. https://doi.org/10.3390/su7044707.

Ridoutt, B. G., Baird, D. and Hendrie, G. A. (2021). Diets within planetary boundaries: What is the potential of dietary change alone?, *Sustainable Production and Consumption* 28, 802–810. https://doi.org/https://https://doi.org/10.1016/j.spc.2021.07.009.

Sallam, A. E., El-feky, M. M. M., Ahmed, M. S. and Mansour, A. (2022) Potential use of whey protein as a partial substitute of fishmeal on growth performance, non-specific immunity and gut histological status of juvenile European seabass, Dicentrarchus labrax, *Aquaculture Research* 53(4), 1527–1541. https://doi.org/10.1111/are.15688.

San Martin, D., Orive, M., Martínez, E., Iñarra, B., Ramos, S., González, N., de Salas, A. G., Vázquez, L. and Zufía, J. (2017) Decision Making Supporting Tool Combining AHP Method with GIS for Implementing Food Waste Valorisation Strategies, *Waste and Biomass Valorization* 8(5), 1555–1567. https://doi.org/10.1007/s12649-017-9976
-z.

Saravani, M., Ehsani, A., Aliakbarlu, J. and Ghasempour, Z. (2019). Gouda cheese spoilage prevention: Biodegradable coating induced by Bunium persicum essential oil and lactoperoxidase system, *Food Science and Nutrition* 7(3), 959–968. https://doi.org
/10.1002/fsn3.888.

Schenker, U., Espinoza-Orias, N. and Popovic, D. (2014). EcodEX: A simplified ecodesign tool to improve the environmental performance of product development in the food industry. In: Proceedings of the 9th International Conference on Life Cycle Assessment in the Agri-Food Sector (LCA Food 2014), San Francisco, California, USA, 8-10 October, 2014. American Center for Life Cycle Assessment. pp. 1220–1223.

Schingoethe, D. J. (1976). Whey utilization in animal feeding: A summary and evaluation, *Journal of Dairy Science* 59(3), 556–570. https://doi.org/jds.S0022-0302(76)84240-3.

Silva, D., Nunes, A. O., da Silva Moris, A., Moro, C. and Piekarski, T. O. R. (2017). How important is the LCA software tool you choose Comparative results from GaBi, openLCA, SimaPro and Umberto. In: Proceedings of the VII Conferencia Internacional de Análisis de Ciclo de Vida en Latinoamérica, Medellin, Colombia. pp. 10–15.

Somlai, R. (2022). Integrating decision support tools into businesses for sustainable development: A paradoxical approach to address the food waste challenge, *Business Strategy and the Environment* 31(4), 1607–1622. https://doi.org/10.1002/bse.2972.

Smithers, G. W. (2008). Whey and whey proteins—from 'gutter-to-gold', *International Dairy Journal* 18(7), 695–704. https://doi.org/10.1016/j.idairyj.2008.03.008.

Sprague, R. H. (1980). A framework for the development of decision support systems, *MIS Quarterly* 4(4), 1–26. https://doi.org/10.2307/248957.

Tait, S., Harris, P. W. and McCabe, B. K. (2021). Biogas recovery by anaerobic digestion of Australian agro-industry waste: A review, *Journal of Cleaner Production* 299, 126876. https://doi.org/10.1016/j.jclepro.2021.126876.

Teigiserova, D. A., Hamelin, L. and Thomsen, M. (2020). Towards transparent valorization of food surplus, waste and loss: Clarifying definitions, food waste hierarchy, and role in the circular economy, *Science of the Total Environment* 706, 136033. https://doi.org/10.1016/j.scitotenv.2019.136033.

Tsermoula, P., Khakimov, B., Nielsen, J. H. and Engelsen, S. B. (2021). Whey - the waste-stream that became more valuable than the food product, *Trends in Food Science and Technology* 118, 230–241. https://doi.org/10.1016/j.tifs.2021.08.025.

Verghese, K., Lockrey, S., Rio, M. and Dwyer, M. (2018). DIRECT, a tool for change: Co-designing resource efficiency in the food supply chain, *Journal of Cleaner Production* 172, 3299–3310. https://doi.org/10.1016/j.jclepro.2017.10.271.

Watson, K. S., Peterson, A. E. and Powell, R. D. (1977). Benefits of spreading whey on agricultural land, *Journal (Water Pollution Control Federation)* 49(1), 24–34. https://www.jstor.org/stable/25039215.

Williams, H., Lindström, A., Trischler, J., Wikström, F. and Rowe, Z. (2020). Avoiding food becoming waste in households – the role of packaging in consumers' practices across different food categories, *Journal of Cleaner Production* 265, 121775. https://doi.org/10.1016/j.jclepro.2020.121775.

WRAP (2022). Food surplus and waste data capture sheet. Available at: https://wrap.org.uk/resources/tool/food-loss-and-waste-data-capture-sheet. WRAP, Banbury, UK.

WRI (2019). Food loss and waste (FLW) value calculator. Available at: http://www.flwprotocol.org/why-measure/food-loss-and-waste-value-calculator/ (Accessed 30 March 2022).

Wu, X., Dong, C., Yao, W. and Zhu, J. (2011). Anaerobic digestion of dairy manure influenced by the waste milk from milking operations, *Journal of Dairy Science* 94(8), 3778–3786. https://doi.org/10.3168/jds.2010-4129.

Zhang, W. and Rhim, J.-W. (2022). Functional edible films/coatings integrated with lactoperoxidase and lysozyme and their application in food preservation, *Food Control* 133, 108670. https://doi.org/10.1016/j.foodcont.2021.108670.

Zotta, T., Solieri, L., Iacumin, L., Picozzi, C. and Gullo, M. (2020). Valorization of cheese whey using microbial fermentations, *Applied Microbiology and Biotechnology* 104(7), 2749–2764. https://doi.org/10.1007/s00253-020-10408-2.

Printed in the USA
CPSIA information can be obtained
at www.ICGtesting.com
JSHW010221021024
70871JS00003B/19